22 Springer Series in Solid-State Sciences

Edited by Manuel Cardona

Springer Series in Solid-State Sciences

Editors: M. Cardona P. Fulde H.-J. Queisser

M. Lannoo   J. Bourgoin

Point Defects in Semiconductors I

Theoretical Aspects

With a Foreword by J. Friedel

With 87 Figures

Springer-Verlag Berlin Heidelberg New York 1981

Dr. *Michel Lannoo*
Dr. *Jacques Bourgoin*

Laboratoire d'Etude des Surfaces et Interfaces, Physique des Solides,
Institut Superieur d'Electronique du Nord, 3, Rue François Base,
F-59046 Lille Cédex, France

Series Editors:

Professor Dr. Manuel Cardona
Professor Dr. Peter Fulde
Professor Dr. Hans-Joachim Queisser

Max-Planck-Institut für Festkörperforschung, Heisenbergstrasse 1
D-7000 Stuttgart 80, Fed. Rep. of Germany

ISBN-13 : 978-3-642-81576-8   e-ISBN-13 : 978-3-642-81574-4
DOI : 10.1007 / 978-3-642-81574-4

Library of Congress Cataloging in Publication Data. Lannoo, M. (Michel), 1942-. Point defect in semiconductors. (Springer series in solid-state sciences ; 22) Includes bibliographical references and index.
Contents: 1. Theoretical aspects. 1. Semiconductors-Defects. 2. Point defects. I. Bourgoin, J. (Jacques),
1938-. II. Title. III. Series.   QC611.6.D4L36   537.6'22   81-5354 AACR2

Offset printing: Beltz Offsetdruck, 6944 Hemsbach/Bergstr. Bookbinding: J. Schäffer oHG, 6718 Grünstadt.
2153/3130-5 4 3 2 1 0

To Ginette and Helma

Foreword

From its early beginning before the war, the field of *semiconductors* has developped as a classical example where the standard approximations of 'band theory' can be safely used to study its interesting electronic properties. Thus in these covalent crystals, the electronic structure is only weakly coupled with the atomic vibrations; one-electron Bloch functions can be used and their energy bands can be accurately computed in the neighborhood of the energy gap between the valence and conduction bands; n and p doping can be obtained by introducing substitutional impurities which only introduce shallow donors and acceptors and can be studied by an effective-mass weak-scattering description.

Yet, even at the beginning, it was known from luminescence studies that these simple concepts failed to describe the various 'deep levels' introduced near the middle of the energy gap by strong localized imperfections. These imperfections not only include some interstitial and many substitutional atoms, but also 'broken bonds' associated with surfaces and interfaces, dislocation cores and 'vacancies', i.e., vacant lattice sites in the crystal. In all these cases, the electronic structure can be strongly correlated with the details of the atomic structure and the atomic motion. Because these 'deep levels' are strongly localised, electron-electron correlations can also play a significant role, and any weak perturbation treatment from the perfect crystal structure obviously fails. Thus, approximate 'strong coupling' techniques must often be used, in line with a more chemical description of bonding.

Now defects that introduce those deep levels are of great technical importance as they regulate many optical and electrical properties of interest as well as being involved in properties useful in the setting up of the electronic devices.

These two volumes concentrate on the field of *point defects*, i.e., impurities and displaced or missing atoms in the crystal lattice. These defects on an atomic scale are the most prevalent. They are also probably the best known and best understood currently. The reader will see, although leading to the 'simplest' questions in the general field of deep levels, the study of point defects already provides ample room for discussing many points both of experimental and theoretical interest. Written by two of the people who have introduced stimulating ideas in the field, these volumes should be of interest both to the specialists and, on a more general level, as a step towards chemistry.

J. Friedel

Preface

This book originates from lectures given at Paris and Lille Universities (in 1977, 1979, and 1980) in the program of "3ème Cycle" attended by graduate students preparing a Ph. D. in Solid State Physics. It is intended for non-specialists, e.g., graduate students who have only an honest background in solid-state physics. For this reason it contains many elementary developments, but it has also been written to be useful for specialists, who are not always familiar with all the different aspects, theoretical as well as experimental, of defect properties in semiconductors.

There are only a few books devoted to the subject of point defects in crystalline semiconductors, and they are research monographs. The reason why textbooks on this subject are not found is that an overall understanding on defects have not yet been achieved in semiconductor materials. However, in the last few years, the situation has improved and we believe that this question can now be treated at a pedagogical level. We have tried to illustrate the theoretical considerations using very simple models, some of which have not been previously published in the literature. We hope that this will encourage more experimentalists to make contact with the basic theory. Though no attempt was made to give an extensive bibliography, references to some basic or review papers where more detailed references can be found have been provided.

This book is divided into two volumes. The first volume primarily treats the theoretical aspects that are necessary for a deeper understanding of the experimental situations, which is mainly described in the second volume. Before giving more details about the content of these volumes, let us first discuss what are the physical properties, associated with the presence of point defects, which are typical of a semiconductor. Intrinsic defects can exist due to a lack of stoichiometry, to thermal treatment, or to thermal agitation created simply by high temperatures. Extrinsic defects, e.g., impurities, can be introduced intentionally or unintentionally during or after its growth. Controlled amounts of Group III or V impurities are introduced in a

Group IV semiconductor to dope it; unwanted impurities (such as carbon and oxygen in silicon, for instance) can diffuse from the environment into the material during the growth process.

The electronic properties of a semiconductor are particularly sensitive to the presence of defects, and some of these properties are sometimes a direct consequence of this presence. In the temperature range where semi-conductor devices are used, the concentration n of the free charged carriers, determined by the amount of doping impurities, is rather small (10^{11}-10^{19} cm^{-3}), and the properties related to the presence of these carriers can be perturbed by a small concentration of defects, typically 10^{-3} n. The reason is that a defect can introduce one or more localized electronic states in the forbidden gap of a semiconductor. A localized state physically corresponds to the fact that the electron wave function decays exponentially from the defect site. A shallow state corresponds to a rather large orbit for the bound elec-tron (typically 10-20 interatomic distances), while a deep state corresponds to a smaller orbit (including typically the first and second neighbors of the defect). Electrons or holes of the shallow states associated with the doping impurities can be easily (e.g., at relatively low temperature) ionized in the conduction or valence bands and provide the free carriers necessary for elec-trical conduction. Deep levels trap these free carriers and therefore reduce their concentration. The carrier mobility is modified because the defects act as scattering centers. The carrier lifetime is modified because the defects levels act as recombination centers; the recombination of electrons with holes from the conduction band to the valence band occurs by the interposition of the deep state. This recombination can be radiative (giving rise to lumines-cence) or nonradiative. Electronic transitions from a localized level to a band can also be induced by light, giving rise to an optical absorption. Finally, defects can exhibit paramagnetic properties; when an unpaired elec-tron bound to a defect site is placed in a magnetic field its level splits and an absorption can be induced between the split levels.

Defects also play a role in all the other physical properties of a semi-conductor. The vibrational modes of the lattice are modified by the presence of defects. The defect locally perturbs the vibrational spectrum and can give rise to localized vibrational modes whose effect is to induce infrared ab-sorption. Defects are also responsible for most of the phenomena associated with atomic transport such as migration, diffusion, precipitation. For in-stance, the jump of a substitutional atom, from one position in the lattice to a neighboring equivalent one, necessitates the presence of an empty site, a vacancy.

This enumeration of defects properties in semiconductors provides the subject matter of this book. First, the definition and geometry, that is the atomic configuration, of the basic defects will be given in order to study their symmetry properties. Next, all the physical properties being dependent on the atomic configuration and on the electronic structure of the defect are treated. Chapter 2 is devoted to the theory of shallow states; this now classical theory is presented in real space, which is more directly understandable than its usual presentation in the reciprocal space. The theory of deep states is developed in two chapters (3 and 4); the first one deals with elementary tightbinding descriptions, the second discusses more sophisticated treatments and the influence of many electron effects. Chapter 5 deals with vibrational modes and entropies, the knowledge of which is necessary to treat atomic transport phenomena. Chapter 6 is devoted to the thermodynamics of defects, e.g., to the question of the nature and of the concentration of the defects which exist at thermal equilibrium. Finally, the last chapter (7) treats atomic transport phenomena, e.g., defect migration and diffusion.

These subjects are presented in the first volume because they can be studied practically independently and are necessary for an understanding of the properties developed in the second volume. These other defects properties are the Jahn Teller effect, optical properties, electron spin resonance, and electrical properties. The two last chapters treat defect creation and annealing kinetics because the study of these processes is a powerful way to obtain information on point defects. Some other properties, such as thermal and mechanical properties, Raman scattering, etc., are not treated because they do not provide direct information on point defects.

Finally, let us mention that we have restricted the book to the case of point defects because we feel that at this time, a unified treatment including one-dimensional defects (dislocations) and two-dimensional defects (surfaces, grain boundaries) is not yet possible.

Lille, February, 1981   *M. Lannoo   J. Bourgoin*

Contents

1. Atomic Configuration of Point Defects

In this first chapter, we define the objects that we shall be dealing with throughout this textbook. The defects are defined by their chemical nature and their geometrical configuration. As will be seen in [1.1], the geometrical configuration, which includes the interaction of the defect with the lattice, i.e., the lattice rearrangement around the defect, can be experimentally obtained from "spectroscopic" measurements (electron paramagnetic resonance and optical techniques). Considerations on defect geometry are necessary from the beginning for two reasons: first, atomic configurations and electronic structures are not independent, and secondly, the symmetry allows one, through the use of group theory, to simplify the treatment of electronic structures.

1.1 Definition of Point Defects

A point defect in a crystal is an entity that causes an interruption in the lattice periodicity. This occurs during the following circumstances.

a) An atom is removed from its regular lattice site; the defect is a vacancy.

b) An atom is in a site different from a regular (substitutional) lattice site; the defect is an interstitial. An interstitial defect can be of the same species as the atoms of the lattice (it is an intrinsic defect, the self-interstitial) or of a different nature (it is then an extrinsic defect, an interstitial impurity).

c) An impurity occupies a substitutional site.

Various kinds of defects are also formed by the association of intrinsic or extrinsic, substitutional or interstitial defects. For instance, a vacancy close to a self-interstitial is a Frenkel pair; two vacancies on neighboring lattice sites form a divacancy, etc. In compound semiconductors all the lattice sites, interstitial as well as substitutional, are not equivalent and a larger variety of intrinsic as well as extrinsic defects exists. For instance, in a 2-6 or 3-5 compound, there are two different sublattices each having its own vacancy, its own interstitial, and its own substitutional or interstitial impurity. An atom of one sublattice placed in the other sublattice forms an "antisite" defect.

All these various types of defects are schematized in Figs.1.1,2. They are "point" defects in contrast to one-dimensional defects (dislocations), two-dimensional defects (surfaces, grain boundaries) or three-dimensional defects (voids, cavities). Small aggregates of several point defects can still be considered as point defects. In that case, the frontier between a point defect and a three-dimensional one is not well defined.

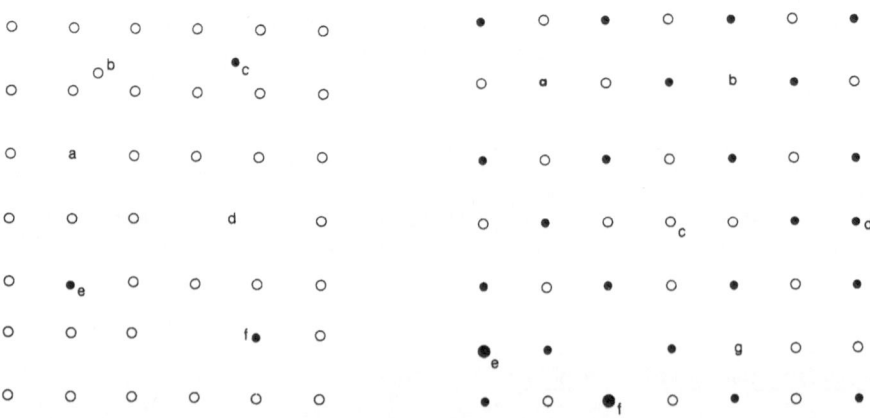

Fig.1.1. Schematic representation of simple point defects in a group IV semiconductor: (a) vacancy; (b) self-interstitial; (c) interstitial impurity; (d) divacancy; (e) substitutional impurity; (f) vacancy-substitutional impurity complex

Fig.1.2. Schematic representation of simple point defects in a compound semiconductor ($\bullet$) atom of sublattice A; (o) atom of sublattice B; (a) vacancy V_A; (b) vacancy V_B; (c) antisite B_A; (d) antisite A_B; (e) substitutional impurity I_B; (f) $I_A + V_B$ complex; (g) $V_B + B_A$ complex

The notion of point defect implies that the perturbation of the lattice remains localized, i.e., it involves an atomic site and few neighbors. But the associated electronic perturbation can extend to larger distances and be, at the limit, delocalized. The above definition of point defects makes reference to a perfect system with translational periodicity which is a result of long-range order. The perfect lattice arrangement is only broken in a localized region. However, the geometrical definition requires only the existence of a short-range order. Consequently, it is possible to conceive the existence of the same point defects in amorphous covalent materials since,

in these materials, the short-range order is preserved. Indeed in amorphous
covalent materials, the interatomic distance as well as the bond angle are
nearly equal to their counterparts in the crystal. Consequently, the envi-
ronments of a point defect in both types of materials, because of the distortion and relaxation the defect induces, are qualitatively similar.

Finally, we note that, when the concentration of the defects is large enough
so that they interact (i.e., there is an overlap of the individual perturbations they induce), they cannot be considered as isolated point defects; defect ordering can occur (alloys).

1.2 Geometrical Configuration of Point Defects

1.2.1 The Vacancy

Four bonds are broken in order to remove an atom from its lattice site and
form a vacancy (Fig.1.3a). The broken (dangling) bonds can form new bonds

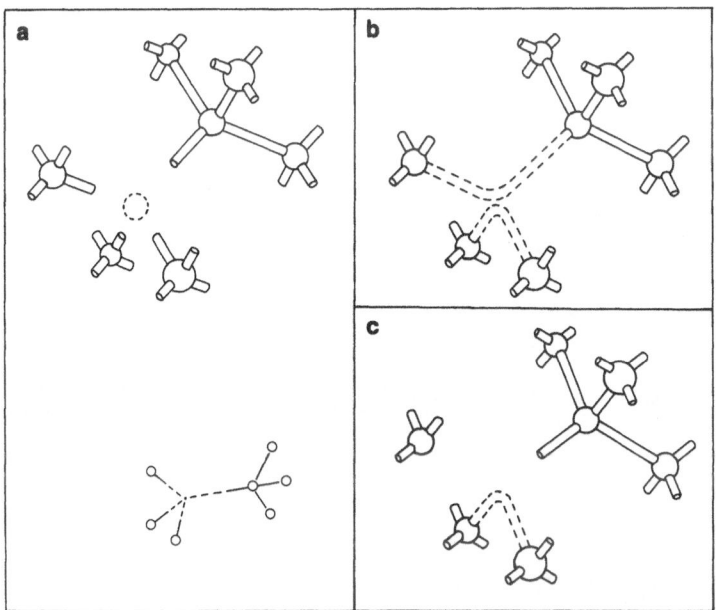

Fig.1.3a-c. The vacancy in diamond lattice and its schematic representation
in two dimensions. (a) Four bonds are broken in order to create the vacancy.
(b) When there is one electron per dangling bond (i.e., for the neutral
vacancy V^0, they form two new bonds leading to local distortion. (c) When an
electron is missing (i.e., for the positive vacancy V^+), one of these two
bonds is weakened since it contains only one electron. The distortion is thus
different from that in the case of V^0

4

leading to atomic displacements. This bonding depends on the charge state
of the vacancy, i.e., on the number of electrons which occupy these dangling
bonds (Figs.1.3b and c). The small atomic displacements of the neighbors of
the vacancy can be inward or outward displacements that preserve the local
symmetry (relaxation) or alter it (distortion). The amplitude of these displacements as well as the new symmetry depend on the type of the bonding,
i.e., on the charge state. Relaxation and distortions will be treated in
Sect.1.3. Here, we consider only one particular case for a possible distortion, one which corresponds to the "split-vacancy" configuration. This special
configuration is important because it is the saddle point configuration of
the normal vacancy during its migration (Chap.7). In this configuration, one
neighbor of the vacancy is displaced half waybetween its original position
and the center of the vacancy (Fig.1.4).

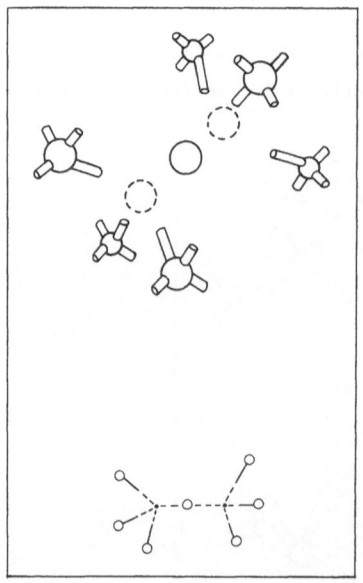

Fig.1.4. The so-called split-vacancy configuration and its schematic two-dimensional
representation

1.2.2 The Divacancy

The divacancy is formed by the removal of two neighboring atoms (Fig.1.5).
The split-divacancy configuration, corresponding to the configuration of the
divacancy in the saddle point for the migration, is given in Fig.1.6. As for
the vacancy case, the type of bonding (and hence the distortion and relaxation
the neighboring atoms undergo) the dangling bonds form is dependent upon their
electronic occupancy.

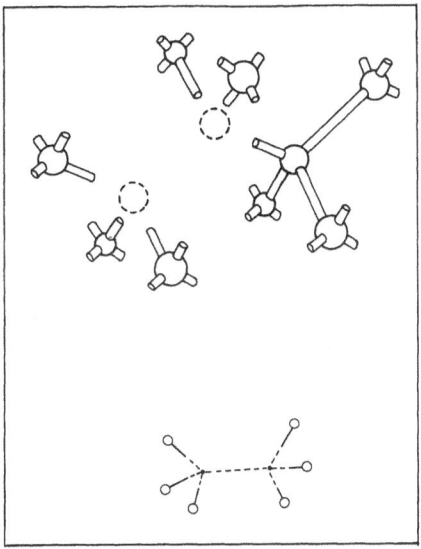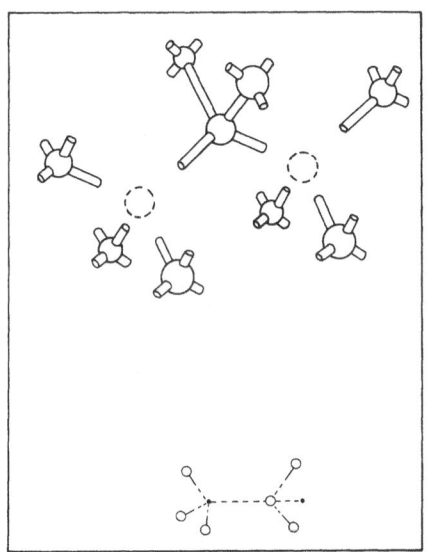

Fig.1.5. The divacancy configuration
and its schematic two-dimensional
representation

Fig.1.6. The split-divacancy configuration and its schematic two-
dimensional representation

1.2.3 The Interstitial

It is impossible to decide a priori what are the stable sites for an interstitial atom. However, we can say that in positions of high symmetry, the
total electronic energy (with all other atoms at their perfect positions)
will be an extremum. It is thus reasonable to consider that some of these
high-symmetry sites are the stable interstitial positions. Because of the
symmetry of the lattice, there may be several equivalent positions per unit
cell.

Two neighboring stable interstitial sites are separated by other high-
symmetry positions which correspond to saddle points of the electronic energy when all other atoms are again kept fixed at their perfect crystal positions. (The migrating path for the interstitial will thus be from one stable
site to another one through this saddle point position.) All these arguments
based on the symmetry of the lattice can be altered, for instance, when the
electron-phonon interaction is taken into account. We shall see later that
this interaction can give rise to distortions of the system. As a result the
stable positions will no longer be those of high symmetry. There can be "off-
centered" configurations in which the interstitial is slightly displaced from

6

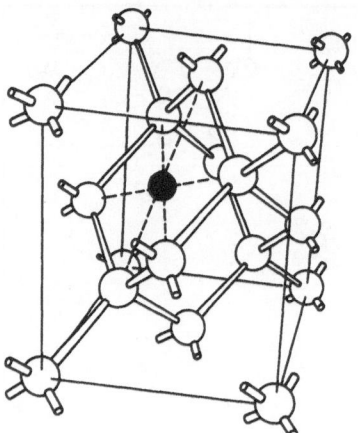 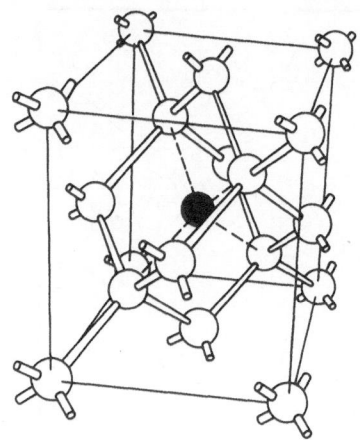

Fig.1.7. The hexagonal interstitial configuration

Fig.1.8. The tetrahedral interstitial configuration

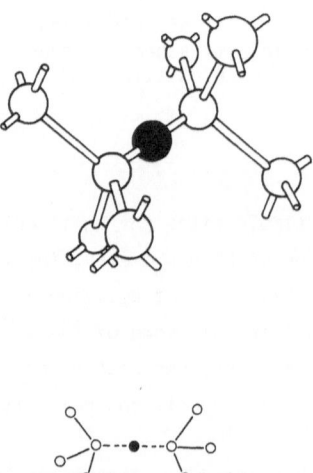

Fig.1.9

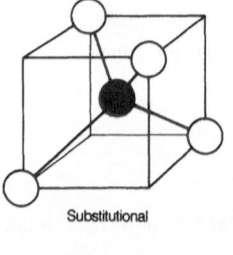 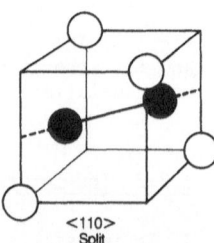

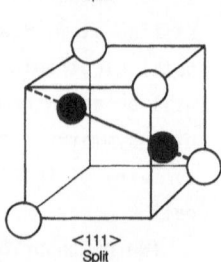

Fig.1.10

Fig.1.9. The bond-centered interstitial configuration and its schematic two-dimensional representation

Fig.1.10. Some split-interstitial configurations

its ideal site. In this respect, once the ideal site has been identified, the situation becomes much the same as for the vacancy.

In the diamond lattice, the sites of highest symmetry are the hexagonal and tetrahedral sites depicted on Figs.1.7,8. Another position of high symmetry corresponds to the "bond-centered" configuration (the interstitial sits at the center of the bond) (Fig.1.9). A possible distorsion of this configuration leads to the "split" interstitial configuration. In this configuration, the interstitial atom and one of its neighbors are displaced in such a way that they form a dumbell centered on the original substitutional site of this neighbor. This split interstitial configuration can have various orientations depending upon the orientation of this dumbell (Fig.1.10).

Once again, the introduction of an interstitial induces a relaxation and a distorsion of the lattice which surrounds it. The type of configuration the interstitial chooses depends on its ability to make bonds with its neighbors and therefore can change with its charge state.

1.2.4 Complex Defects

When a simple defect moves, it can interact with other intrinsic as well as extrinsic point defects giving rise to a more complex defect. For instance, when the vacancy becomes mobile in silicon, it can be trapped by an oxygen impurity (present in Czochralski grown material) and form a V - 0 complex (the A center), or by the doping impurity (Al for instance) and form a V - Al complex (the E center), or by another vacancy (in undoped floating zone material) and form divacancies. Figure 1.11 gives the configuration of the A center in which the oxygen atom occupies a position slightly displaced from the substitutional vacancy site.

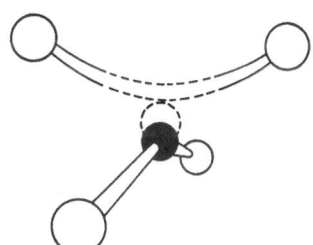

Fig.1.11. In the A-center configuration (vacancy + oxygen complex), the oxygen atom is slightly displaced off the substitutional position

8

1.2.5 Aggregates

When the concentration of a particular defect is large, they tend to aggregate as the temperature is increased. Vacancies form divacancies that upon becoming mobile or dissociating, form trivacancies, quadrivacancies and so on. In principle, the larger the number of defects involved in an aggregate, the larger the number of possible configurations (e.g., Fig.1.12 which gives one of the possible configurations for a penta-vacancy). But when the número of vacancies in the complex is large, they tend to arrange themselves in lines, rings, or platelets. This behavior should also be true for self-interstitials and for any type of extrinsic defects.

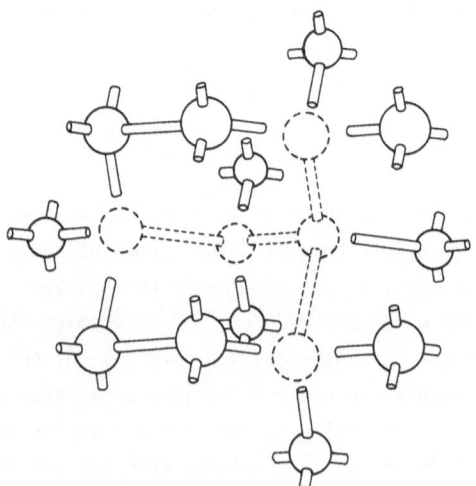

Fig.1.12. One of the possible configurations for a penta-vacancy

1.3 Lattice Distortion and Relaxation

We have already mentioned that the introduction of a point defect induces displacements of the lattice atoms which surround it. The atoms involved are first neighbors, second neigbors, etc., depending on the range of the perturbation introduced by the defect. When the symmetry of the lattice is conserved, these displacements are said to result in a relaxation; when the symmetry is lowered, the induced displacements are said to result in a distortion. The amplitude as well as the type of displacement the lattice atoms

undergo is obtained theoretically by minimizing the total energy of the system, lattice plus defect, versus the positions of the various atoms involved
in the distortion or in the relaxation. The way this total energy is obtained
is described in Chap.4; in Chap. 6 the formation energy of the vacancy is
calculated taking into account the distortion. With an increasing number of
atoms involved in the distortion, the number of displacements to be considered
increases rapidly and the problem becomes quickly impracticable. Actually, as
it will be discussed in Chaps.4 and 6, the problem of the evaluation of a
total energy is not yet solved.

The type of distortion depends on the way the defect is bonded to the
neighboring atoms. As a result, the distortion is a function of the charge
state of the defect. This notion is very important, because it results in a
charge-state dependence of energies and entropies. Consequently all the properties which are related to these quantities, defect concentration at thermal
equilibrium, stability, migration, diffusion, solubility, vibrational modes,
electron-phonon interaction, etc. will be charge-state dependent. A manifes-
tation of this charge state dependence of a distortion is the Jahn-Teller
effect. The Jahn-Teller effect is the subject of a particular chapter [Ref.
1.1, Chap.8] and its application to the case of the vacancy will be treated
there in detail.

1.4 Defect Symmetry and Group Theory

Crystal lattices have, in general, symmetry properties. This means there
exist symmetry operations R (such as rotations, inversions ...) around one
of the normal lattice sites which leave the crystal unchanged. In other words,
any atom of the crystal after a symmetry operation is replaced by an equivalent one. Consider, for instance, the local situation around one atom O in
the diamond lattice. Figure 1.13b is obtained from Fig.1.13a by a rotation
of π around the axis $\vec{Ox}$. Since all atoms are identical, both situations are
exactly identical. This rotation is then a symmetry operation. There are
other such operations. It can be shown that all of them form a group, which
in this example is labelled T_d, the tetrahedral point group. The point group
defines completely the symmetry properties of a lattice around one of its
sites.

Beyond its interest as regards to the classification of each lattice (or
molecule) with respect to its symmetry properties, group theory can be very

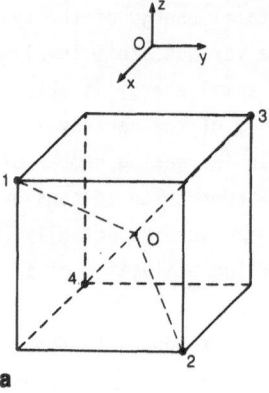
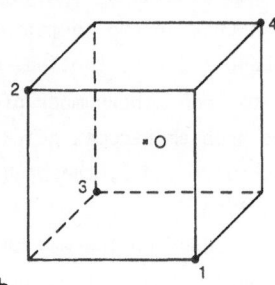

Fig.1.13a,b. Local atomic coordination of the diamond lattice. A rotation of π around $\vec{Ox}$ results in atomic changes from case (a) to case (b)

helpful in simplifying the resolution of quantum mechanical problems[1]. Well-known examples are the hydrogen molecule and Bloch's theorem. In the case of the H_2 molecule, the eigenfunctions are classified as symmetrical or anti-symmetrical. This allows a direct decomposition of the Hamiltonian matrix into two independent parts, corresponding to these two types of solutions. In the second case the translational symmetry is exploited. The wave functions are classified with respect to the wave vector. The Hamiltonian matrix now factorizes into N independent blocks, where N is the number of unit cells of the crystal.

In this section we give a brief account of the application of group theory with a few examples for defect problems. We first show how the Hamiltonian matrix can be factorized by symmetry considerations.

1.4.1 Factorization of the Hamiltonian

The basic principles of the factorization come from the properties of commuting operators. Assume that the Hamiltonian H commutes with another operator Γ. Expressing the operators under matrix form in a given basis, the condition of commutation is written

$$(\Gamma H)_{ij} = (H\Gamma)_{ij} \tag{1.1}$$

---

[1] For a simple introduction to group theory see [1.2]. Classical treatises of group theory are found in [1.3,4], and applications to molecular systems in [1.5] and to crystallography in [1.6].

or

$$\sum_k \Gamma_{ik} H_{kj} = \sum_k H_{ik} \Gamma_{kj} \quad . \tag{1.2}$$

We choose a basis which diagonalizes the operator Γ. Then (1.2) reduces to

$$(\Gamma_{ii} - \Gamma_{jj}) H_{ij} = 0 \quad . \tag{1.3}$$

This means that if Γ_{ii} is different from Γ_{jj}, then H_{ij} is zero. In other words, the matrix H is automatically factorized into blocks, each of them corresponding to a given subset of eigenstates of Γ leading to the same eigenvalue.

To relate this to the symmetry properties of the system, let us consider all possible independent symmetry operations R. They form a group. In many cases it is interesting to associate to each operation a matrix $\Gamma(R)$ whose size does not depend on R (we shall give examples below). These matrices form a representation of the group. Their advantage is that general theorems exist that allow their decomposition into independent blocks. If, moreover, they commute with H then, from (1.3), the decomposition of H is also achieved.

Usually we have to diagonalize a Hamiltonian matrix using a finite set of basis functions $\phi_i(\underline{x})$ where $\underline{x}$ is the set of coordinates. Any symmetry operation R acting on this set gives a new one $\underline{x}'$ given by

$$\underline{x}' = R\underline{x} \quad , \tag{1.4}$$

which under matrix form gives

$$x_i' = \sum_j R_{ij} x_j \quad , \tag{1.5}$$

x_i' and x_j being coordinates along the ith and j^{th} unit vector. The set of matrices R_{ij} for the different operations R forms a representation of the group which may be useful in some cases. For our purposes, it is interesting to introduce new operators P_R associated to R which operate on functions rather than on coordinates. To define the action of P_R on any function $f(\underline{x})$, a natural convention is to write

$$P_R f(R\underline{x}) = f(\underline{x}) \quad , \tag{1.6}$$

which means that transforming f and $\underline{x}$ together by the operation R does not change anything. This convention becomes evident in Fig.1.14 where the function f is an atomic orbital centered on a nucleus. If both the orbital and the system of coordinates are rotated by θ, then the displaced orbital has

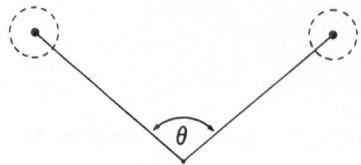

Fig.1.14. Effect of a rotation of angle θ around an axis on a nucleus and its associated orbital

exactly the same expression in the new system of coordinates as the original orbital in the initial system. The mathematical definition (1.6) of P_R is usually written as

$$P_R f(\underline{x}) = f(R^{-1}\underline{x}) \quad . \tag{1.7}$$

The set of different P_R form a group isomorphic to the operations R and thus have exactly corresponding properties.

Let us then consider basis functions $\phi_j(\underline{x})$ such that the action of P_R on a function $\phi_i(x)$ gives a function that can be expressed in the same basis set. This means that P_R transforms the basis functions $\phi_i(\underline{x})$ between themselves, i.e.,

$$P_R \phi_i(\underline{x}) = \sum_j (P_R)_{ij} \phi_j(\underline{x}) \quad . \tag{1.8}$$

The matrices $(P_R)_{ij}$ again form a representation of the group. They correspond to symmetry operations which leave the system invariant. Thus the action of $P_R H$ on a function $f(\underline{x})$ [which gives $P_R Hf(\underline{x})$] is identical to $HP_R f(\underline{x})$, i.e., the two operators commute. The matrices $(P_R)_{ij}$ therefore provide representations Γ which, by virtue of (1.3), can help in factorizing the Hamiltonian matrix. The whole problem now consits in factorizing such a representation Γ(R). Group theory provides general theorems that allow this to be done in a systematic manner.

1.4.2 Irreducible Representations

Let us now investigate how the matrix Γ(R) can be put under block form, i.e., written in the form

$$\Gamma(R) = \begin{bmatrix} \Gamma^{(1)}(R) & 0 & 0 \\ 0 & \Gamma^{(2)}(R) & 0 \\ 0 & 0 & --- \end{bmatrix} \quad . \tag{1.9}$$

When this can be done by a suitable basis change, the representation $\Gamma(R)$ is reducible. When this cannot be done, the representation is irreducible. When the decomposition of $\Gamma(R)$ in (1.9) is completed, it is evident that the blocks $\Gamma^{(i)}(R)$ are irreducible representations. We can thus formally write (1.9) as follows

$$\Gamma(R) = \sum_i a_i \Gamma^{(i)}(R) \quad . \tag{1.10}$$

This expresses the fact that $\Gamma(R)$ can be reduced into the irreducible representations $\Gamma^{(i)}(R)$, each of them having the possibility of ocurring a_i times.

There are a number of theorems to find this decomposition. The first of them (the great orthogonality theorem) leads to

$$\sum_R \Gamma^{*(i)}_{\mu\nu}(R)\Gamma^{(j)}_{\alpha\beta}(R) = \frac{h}{\ell_i}\, \delta_{ij}\delta_{\mu\alpha}\delta_{\nu\beta} \quad , \tag{1.11}$$

where h is the order of the group (the number of distinct operations R) and ℓ_i the dimensionality of the matrix $\Gamma^{(i)}(R)$. It can be shown that this theorem implies that

$$\sum_i \ell_i^2 = h \quad . \tag{1.12}$$

For many purposes the great orthogonality theorem contains too much information. We can condense it by using the trace of a given irreducible representation,

$$\chi^{(i)}(R) = \text{Trace}\{\Gamma^{(i)}(R)\} \quad , \tag{1.13}$$

which is called the character of this representation. A straightforward application of (1.11) gives

$$\sum_R \chi^{*(i)}(R)\chi^{(j)}(R) = h\delta_{ij} \quad . \tag{1.14}$$

The elements R of the group can be divided into classes. For instance, in the tetrahedral group T_d, the three rotations of π around the $O\vec{x}$, $O\vec{y}$, $O\vec{z}$ axes belong to the same class. This is also the case for the eight rotations of $2\pi/3$, $4\pi/3$ around the <111> axes. Because the character $\chi^{(i)}(R)$ is the same for all elements R belonging to the same class and if we label τ_k the kth class containing N_k elements, then (1.14) condenses into

$$\sum_k \chi^{*(i)}(\tau_k)\chi^{(j)}(\tau_k) N_k = h\delta_{ij} \quad , \tag{1.15}$$

where $\chi^{(i)}(\tau_k)$ is the common value of the character for all elements belonging to τ_k. Another very important theorem says that the number of different irreducible representations is equal to the number of classes. This number is usually small so that any matrix $\Gamma(R)$ decomposes into a small number of irreducible blocks $\Gamma^{(i)}(R)$, each of them appearing a_i times.

We shall now demonstrate that these theorems allow determination of the number of times a_i that a given $\Gamma^{(i)}(R)$ occurs in the decomposition of $\Gamma(R)$. For this, we take the trace $\chi(R)$ of $\Gamma(R)$ (the character of this representation) which can be written, using expression (1.10), as

$$\chi(R) = \sum_i a_i \chi^{(i)}(R) \tag{1.16}$$

or

$$\chi(\tau_k) = \sum_i a_i \chi^{(i)}(\tau_k) \quad . \tag{1.17}$$

This expression, multiplied by $\chi^{*(j)}(\tau_k)N_k$ and summed over k, leads by virtue of (1.15) to the following result

$$a_j = \frac{1}{h} \sum_k N_k \chi^{*(j)}(\tau_k)\chi(\tau_k) \quad . \tag{1.18}$$

The decomposition of any representation $\Gamma(R)$ can be obtained systematically. The first step is to calculate its character χ or trace for any class τ_k. The second step is to apply (1.18), which directly gives a_j.

It is necessary to know first the characters $\chi^{(j)}(\tau_k)$ of the irreducible representations. This can be determined for any given group, the result being a table of characters (given in any textbook on group theory) which takes the form of Table 1.1.

Table 1.1. Character table of a point group

Group label	τ_1	τ_2	-----	τ_n
$\Gamma^{(1)}$	$\chi^{(1)}(\tau_1)$	$\chi^{(1)}(\tau_2)$	-----	$\chi^{(1)}(\tau_n)$
$\Gamma^{(2)}$	$\chi^{(2)}(\tau_1)$	$\chi^{(2)}(\tau_2)$	-----	$\chi^{(2)}(\tau_n)$
$\vdots$	-------	-------	-----	
$\Gamma^{(n)}$	$\chi^{(n)}(\tau_1)$	$\chi^{(n)}(\tau_2)$	-----	$\chi^{(n)}(\tau_n)$

We do not discuss the methods used to derive this tables of characters here. Instead we detail an example of its application in one important case, the vacancy in a covalent system.

1.4.3 The Example of the Vacancy in the Diamond Lattice

The vacancy is obtained by removing an atom (the central atom 0 in Fig.1.13). The symmetry group, the tetrahedral point group T_d, remains unchanged. For this group the characters [1.13] are given in Table 1.2.

Table 1.2. Character table of T_d group

T_d		E	8 C_3	3 C_2	6 σ_d	6 S_4
$\frac{1}{\sqrt{3}}$ $(x^2+y^2+z^2)$	A_1	1	1	1	1	1
	A_2	1	1	1	-1	-1
$\frac{1}{\sqrt{6}}$ $(3z^2-r^2)$ $\frac{1}{\sqrt{2}}$ (x^2-y^2)	E	2	-1	2	0	0
R_x, R_y, R_z	T_1	3	0	-1	-1	1
x, y, z	T_2	3	0	-1	1	-1

The labelling of the classes begins by E, the identity operation. There are then eight rotations C_3 by $\pm 2\pi/3$ (order 3) around the (111) axes. There are three rotations of order 2 around $O\vec{x}$, $O\vec{y}$, $O\vec{z}$. There are six diagonal reflection planes σ_d and six improper rotations S_4 of which no details shall be given. The total gives five classes and 24 elements. As a result there are five irreducible representations labelled here A_1, A_2, E, T_1, T_2. Their dimensionality is given by the numbers in first column, which are the values of the character for the identity operation E. The irreducible representation for E are unit matrices, so their trace (or character) is equal to their dimensionality. On the same table are also given some simple functions characterizing these irreducible representations (this is discussed later). Let us assume that four equivalent functions $\phi_i(\underline{x})$ centered on the neighbors of the vacancy (Fig.1.13, i = 1 to 4) are sufficient to describe the eigenfunctions of the Hamiltonian H of the vacancy (this is justified in Chap.3). Any symmetry operation obviously transforms these functions into one another. The corresponding matrix $(P_R)_{ij}$ has a dimensionality equal to four and is our reducible representation. Let us then obtain its decomposition.

First of all, we have to calculate its character $\chi(E)$, the trace of that matrix for the identity operator. Each $\phi_i(\underline{x})$ remains unchanged, so that we have a 4×4 unit matrix, leading to $\chi(E) = 4$. To calculate $\chi(C_3)$, it is enough to consider one of the rotations of order 3. Let us then consider the rotation of $2\pi/3$ around the axis joining atom 1 to the origin 0. ϕ_1 remains unaltered while ϕ_2, ϕ_3 and ϕ_4 undergo a cyclic permutation. The diagonal of the 4×4 matrix has one element equal to unity, the other three being zero, and $\chi(C_3)$ = 1. A C_2 operation is, for instance, a rotation of π around $O\vec{x}$. All $\phi_i(\underline{x})$ are changed and $\chi(C_2)$ = 0. The σ_d are diagonal reflection planes, normal to a cube face, passing through a plane of atoms, for instance, atoms 1,2. The two functions ϕ_1 and ϕ_2 are unchanged and $\chi(\sigma_d)$ = 2. The six S_4 are rotations by $\pm \pi/2$ around $O\vec{x}$, $O\vec{y}$, $O\vec{z}$, followed by a reflection in a perpendicular plane passing through the origin. Clearly $\chi(S_4)$ = 0. The character table for this 4×4 representation is given by Table 1.3.

Table 1.3. Character table of the representation formed by the four functions centered on the neighbors of the vacancy

	E	8 C_3	3 C_2	6 σ_d	6 S_4
$\chi(\tau_k)$	4	1	0	2	0

It is now easy to calculate the decomposition of this representation into the irreducible representations. Using (1.18) we obtain

$$a_{A_1} = 1 \quad , \quad a_{A_2} = 0$$

$$a_E = 0 \quad , \quad a_{T_1} = 0 \quad , \quad a_{T_2} = 1 \quad , \tag{1.19}$$

so that we can write the formal equality

$$\Gamma = A_1 + T_2 \quad , \tag{1.20}$$

which means that Γ can be factorized into a 1×1 and a 3×3 blocks. Equation (1.3) tells us that the basis which factorizes Γ will automatically factorize the 4×4 Hamiltonian matrix. We shall show how to construct the corresponding basis functions. But first, we shall discuss several aspects of group theory that are of current practical interest.

1.4.4 Various Points of Interest

One of the first situations that is often encountered when solving the Schrödinger equation is the following. Assume an eigenvalue E_n with different possible eigenstates $\psi_n^{(\alpha)}$. Since any operator P_R defined previously commutes with H, we have

$$P_R H \psi_n^{(\alpha)} = E_n P_R \psi_n^{(\alpha)} \tag{1.21}$$

$$P_R H \psi_n^{(\alpha)} = H P_R \psi_n^{(\alpha)} \quad . \tag{1.22}$$

This means $P_R \psi_n^{(\alpha)}$ belongs to the subset of eigenstates corresponding to the eigenvalue E_n. Thus, the action of P_R on any $\psi_n^{(\alpha)}$ gives a combination of some other members of the subset, and from one $\psi_n^{(\alpha)}$ it is possible to generate other $\psi_n^{(\alpha)}$ by the action of the operators P_R. If this procedure gives all the members of the subset, the degeneracy of E_n is said to be normal, i.e., results from symmetry. If this is not the case, the degeneracy is accidental. In the first case, the degeneracy of E_n is equal to the dimensionality of the irreducible representation generated by the action of P_R on the $\psi_n^{(\alpha)}$.

A consequence is that if the representation corresponding to the matrices of P_R within the basis set can be decomposed with all a_i equal to 0 or to 1, then H is completely factorized. Irreducible representations of dimensionality greater than one will correspond to energy levels with identical degeneracy. If an irreducible representation of dimensionality n corresponds to a value of a_i equal to p, then the corresponding block in H is of size np × np. The diagonalization of such a block will necessarily yield p distinct levels, with a degeneracy equal to n.

A second important situation corresponds to a representation that is said to be the direct product of other representations. This occurs, for instance, when we use as basis functions products of the basis functions corresponding to two given representations. If the two bases are ϕ_i (i = 1 to n) and ψ_k (k = 1 to m), then the chosen basis states are the $\phi_i \psi_k$. The action of P_R on such products gives the matrix elements of $\Gamma(R)$, while the action of P_R on the ϕ_i gives $\Gamma_1(R)$ and the action of P_R on the ψ_k gives $\Gamma_2(R)$. We have

$$P_R\phi_i\psi_k = \sum_{i'k'} (\Gamma(R))_{ik,i'k'}\phi_{i'}\psi_{k'} \quad , \tag{1.23}$$

and also

$$P_R\phi_i\psi_k = \sum_{i'} (\Gamma_1(R))_{ii'}\phi_{i'} \sum_{k'} (\Gamma_2(R))_{kk'}\psi_{k'} \quad . \tag{1.24}$$

As a result

$$(\Gamma(R))_{ik,i'k'} = (\Gamma_1(R))_{ii'}(\Gamma_2(R))_{kk'} \quad , \tag{1.25}$$

which we write formally as

$$\Gamma(R) = \Gamma_1(R) \times \Gamma_2(R) \quad . \tag{1.26}$$

The practical problem is that usually we know the decomposition of Γ_1 and Γ_2 and search for the decomposition of $\Gamma(R)$. We then have to first calculate the character $\chi(R)$, i.e., the trace of Γ. From (1.25) it is obvious that

$$\chi(R) = \chi_1(R) \cdot \chi_2(R) \quad . \tag{1.27}$$

Then the analysis proceeds as we have discussed before.

The last comment, which is useful in many cases, corresponds to a lowering of the symmetry with respect to the original symmetry. This can either be produced by some external perturbation, such as an electric or magnetic field, uniaxial stress, etc. (Sect.1.5.1); or by a spontaneous distortion near the defect site, such as the Jahn-Teller effect [Ref.1.1, Chap.8]. Then, the new symmetry group is a subgroup of the original one (it contains only part of its elements). Generally speaking, there are less irreducible representations and they have a lower dimensionality. This means that a lower symmetry decreases the degree of direct factorization. The irreducible representations of the original group now become reducible and can be decomposed into those of the new group. This is tabulated in many textbooks, but it is not difficult to do it directly since we already know the characters of these representations.

Here, we shall give a few details for the two subgroups D_{2d} and C_{3v} of the tetrahedral point group T_d. These correspond, for instance, to distortions around a point defect, lowering the local symmetry to a tetragonal or a trigonal one, respectively. An illustration of these two cases is given in Fig.1.15a for D_{2d} and Fig.1.15b for C_{3v}, where the lowering of symmetry

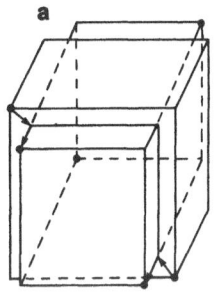 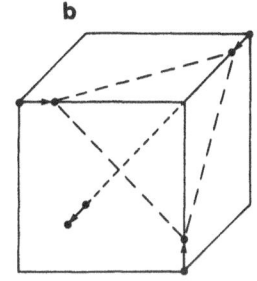

Fig.1.15a,b. Representations of the atomic displacements, the four neighbors
of a substitutional position undergo, leading to: (a) a tetrahedral distortion (D_{2d} symmetry) and (b) a trigonal distortion (C_{3v} symmetry)

results from displacements of the nearest neighbors of the vacancy. The character tables of D_{2d} and C_{3v} are of interest if we want the decomposition
of representations into irreducible components for these subgroups. They are
given in Tables 1.4,5.

Table 1.4. Character table of the
D_{2d} group

D_{2d}	E	C_2	$2C_2'$	$2\sigma_d$	$2S_4$
A_1	1	1	1	1	1
A_2	1	1	-1	-1	1
B_1	1	1	1	-1	-1
B_2	1	1	-1	1	-1
E	2	-2	0	0	0

Table 1.5. Character table of the
C_{3v} group

C_{3v}	E	$2C_3$	$3\sigma_v$
A_1	1	1	1
A_2	1	1	-1
E	2	-1	0

The labelling of the symmetry operations is the same as for T_d except that
for D_{2d}, there are now two sorts of rotations of order 2, C_2 and C_2', and for
C_{3v}, the σ_v are reflection planes containing the axis of rotation, becoming
σ_d in the T_d group.

A special example of great interest is the decomposition of the irreducible representations of T_d into those of D_{2d} and C_{3v}. From (1.18) and Tables
1.2,4,5 we find for a lowering from T_d to D_{2d}:

$$A_1 \rightarrow A_1$$
$$A_2 \rightarrow B_1$$
$$E \rightarrow A_1 + B_1$$
$$T_2 \rightarrow B_2 + E$$
$$T_1 \rightarrow A_2 + E \quad , \tag{1.28}$$

for a lowering from T_d to C_{3v}:

$$A_1 \rightarrow A_1$$
$$A_2 \rightarrow A_2$$
$$E \rightarrow E$$
$$T_2 \rightarrow A_1 + E$$
$$T_1 \rightarrow A_2 + E \tag{1.29}$$

1.4.5 Basis Functions for Irreducible Representations

All we have described up to now leads only to the knowledge of the size of the blocks occurring in the factorization. However, it is also necessary to know the basis functions corresponding to these submatrices. In simple cases, this can practically be done by inspection. In the general case, there exist systematic methods based on the great orthogonality theorem, which we have not used completely.

There exist an infinite number of ways of writing the matrix $\Gamma^{(i)}(R)$ of the ith irreducible representation, corresponding to all possible unitary transformations of its basis states. Usually a standard matrix corresponding to simple basis functions is defined. Such choices are indicated in Table 1.2. For instance, the simple functions x,y,z generate one form of the matrix $\Gamma^{(T_2)}(R)$ which we can use as a reference for all calculations. This means that all functions that we shall find as basis functions of this representation will behave as the simpler functions x,y,z under symmetry operations.

Let us first consider the vacancy case already discussed in Sect.1.4.3. This is a simple example where the basis functions can be easily written by inspection. We have four equivalent functions ϕ_1, ϕ_2, ϕ_3, ϕ_4 forming a 4×4 representation which decomposes into A_1 (of dimensionality one) and T_2 (dimensionality 3). Table 1.2 mentions that $x^2 + y^2 + z^2$ is a simple basis function of A_1. This function is invariant under all symmetry operations. The combination of the ϕ_α which presents this property is obviously

$$\psi_{A_1} = \frac{1}{2} (\phi_1 + \phi_2 + \phi_3 + \phi_4) \quad . \tag{1.30}$$

The irreducible representation T_2 is shown on this table to correspond to the simple functions x,y,z. The combinations of the ϕ_α which have this property are

$$\psi_x = \frac{1}{2} (\phi_1 + \phi_2 - \phi_3 - \phi_4) \quad ,$$

$$\psi_y = \frac{1}{2} (-\phi_1 + \phi_2 + \phi_3 - \phi_4) \quad ,$$

$$\psi_z = \frac{1}{2} (\phi_1 - \phi_2 + \phi_3 - \phi_4) \quad . \tag{1.31}$$

(They have an axial symmetry along $\vec{Ox}$, $\vec{Oy}$, $\vec{Oz}$ as can be seen in Fig.1.13a.)

A systematic procedure to obtain these functions exists, based on the knowledge of the standard matrices $\Gamma^{(i)}(R)$. Always in the same example, the standard matrices $\Gamma^{(T_2)}(R)$ are obtained from the transformation properties of x,y,z and can be either calculated or found in existing tables. Now, given a set of basis functions ϕ_α, we need to find among them what are the basis functions for each irreducible representation. For this, define an operator

$$P_{\lambda K}^{(i)} = \frac{\ell_i}{h} \sum_R \Gamma^{(i)}(R)_{\lambda K}^* P_R \quad , \tag{1.32}$$

where the $\Gamma^{(i)}(R)_{\lambda K}$ are the known elements of the standard irreducible representations. We first use the operator corresponding to $\lambda = K$, i.e., $P_{KK}^{(i)}$. Its action on any ϕ_α leads necessarily, by virtue of the great orthogonality theorem, to a function which has the property of being a Kth basis function of the ith irreducible representation. For instance, $P_{xx}^{(T_2)}$ acting on the function ϕ_1 for the vacancy will lead to the function ψ_x automatically. This can be done by the action of all $P_{KK}^{(i)}$ when varying i and K on all basis functions. If the results are ambiguous, they have to be checked by the action of $P_{\lambda K}^{(i)}$. This operator, acting on any ϕ_α, selects in ϕ_α its component on the Kth basis function and transforms it into the λth basis function. This can be used as a check or to generate other basis functions from a known one.

1.4.6 Simplification of Matrix Elements by Symmetry

In many situations of interest we have to investigate under what conditions a product

$$\phi_{\Gamma^{(i)},m} \times \psi_{\Gamma^{(j)},\ell}$$

of basis functions of rank m and ℓ of two irreducible representations $\Gamma^{(i)}$ and $\Gamma^{(j)}$ is invariant under all symmetry operations, i.e., belongs to A_1. This can be done by projecting this product on A_1 using the operator P^{A1} defined in (1.32) which turns out to be proportional to $\sum_R P_R$. Using then (1.24) and the orthogonality theorem (1.11) we easily find that this can only occur when i is equal to j and m is equal to ℓ. This means that only products such as

$$\phi^*_{\Gamma}(i)_{,m} \psi_{\Gamma}(i)_{,m}$$

contain a part which is invariant. Furthermore, in that case, we have

$$P^{A_1} \phi^*_{\Gamma}(i)_{,m} \psi_{\Gamma}(i)_{,m} = \frac{1}{\ell_i} \sum_{m'} \phi^*_{\Gamma}(i)_{,m'} \psi_{\Gamma}(i)_{,m'} \qquad (1.33)$$

which means that this component is the same whatever the value of m.

This can be generalized to the simplification of matrix elements by symmetry. All of them can be expressed in terms of the integrals of products of basis functions such as

$$\phi^*_{\Gamma}(i)_{,m} \psi_{\Gamma}(j)_{,\ell} \cdot$$

On the other hand any such integral is a scalar quantity and is thus invariant under all symmetry operations. Thus the product $\phi^*\psi$ itself must be invariant. This naturally leads to the relationship

$$\int \phi^*_{\Gamma}(i)_{,m} \psi_{\Gamma}(j)_{,\ell} \, d\tau = \delta_{i,j}\delta_{m,\ell} \left\langle \phi|\psi \right\rangle_{\Gamma}(i) \quad , \qquad (1.34)$$

where $\langle\phi|\psi\rangle_{\Gamma}(i)$ is a shorthand notation, showing that the result is dependent upon the irreducible representation.

1.5 Experimental Determination of Defect Symmetry

A simple defect such as the vacancy, which we have described previously, possesses a given symmetry. Of course, when the static lattice distortion around the defect is included, this symmetry is lowered. The symmetry group of the defect only contains part of the elements of the original point group corresponding to the undistorted situation: it is a subgroup of the original point group. Consider, for instance, the vacancy described in Sect.1.2.1; it possesses the full symmetry of the crystal, i.e., it belongs to the T_d group of the diamond lattice whose elements are given in Sect.1.4.3. When the first neighbors of the vacancy undergo a trigonal or a tetragonal distortion (Fig. 1.15) the symmetry group is reduced to C_{3v} and D_{2d}, respectively (Sect.1.4.4). In this section we shall describe the experimental ways used to determine the symmetry of a defect and consequently the type of distortion it produces. These "spectroscopic" techniques consist in the study of the response of an anisotropic property of the defects, which reflects some elements of its symmetry group, to a "polarized" excitation. Such excitations can be of mechanical origin (a uniaxial stress), a magnetic or an electric field, or polarized light. The response is usually detected through the interaction of the defect with an electromagnetic wave, but electrical properties could also be used in principle. The Hamiltonian δH corresponding to the excitation is usually small and can be treated as a perturbation on the Hamiltonian H of the defect. The matrix elements $<\psi_i|\delta H|\psi_j>$ occurring in first-order perturbation theory will or will not vanish depending on the symmetry properties of the wave functions and the polarized excitation. Depending on the detailed situation (the presence or lack of orbital degeneracy), a splitting or a shift of the electron energy levels is obtained. In addition, the orientational degeneracy can be lifted, at least partially, since the local symmetry of the defect is lowered by the perturbation. It is this splitting of degenerate levels that provides informations on defect symmetry.

The amount of the shift of the electronic level is a measure of the amplitude of the distortion (the amplitude of an atomic displacement is usually given in percentage of the interatomic distance). Unfortunately, this measure necessitates the detailed knowledge of the perturbation δH, which most of the time we do not have. In the same way, the amplitude of a lattice relaxation (which does not modify the symmetry of the defect) is difficult to obtain. In simple cases (of impurities very large compared to the host atoms, for instance), this lattice relaxation can be estimated from the measurements of the local strain field the defects induce using X-ray techniques [1.7]. We shall also see that strong distortions or relaxations induce large entropy terms (Chap.5) and correspond to strong electron-phonon effects (see [Ref. 1.1, Chaps.10,12]), but the measurement of such quantities still cannot provide quantitative informations on this amplitude.

1.5.1 Splitting Under Uniaxial Stress

a) General Considerations

The application of a uniaxial stress on a crystal induces a deformation, i.e., atomic displacements [1.8,9]. Because these displacements are small, the re-

sulting perturbation potential V in the vicinity of the defect can be expressed as a linear function of the displacements, that is of the applied stress
(Hooke's law). We can therefore write

$$V = \sum_{i,k} V_{ik}\sigma_{ik} \quad . \tag{1.35}$$

The σ_{ik} are the elements of a second-rank tensor representing the stress. An
element σ_{xy} is the component of the force per unit surface along the axis x,
applied on the face of the crystal perpendicular to the y axis. In a cubic
crystal the stress tensor takes therefore the following forms,

$$\begin{vmatrix} \sigma_{xx} & \sigma_{xy} & \sigma_{xz} \\ \sigma_{xy} & \sigma_{yy} & \sigma_{yz} \\ \sigma_{xz} & \sigma_{yz} & \sigma_{zz} \end{vmatrix} = \sigma \begin{vmatrix} 1/2 & 1/2 & 0 \\ 1/2 & 1/2 & 0 \\ 0 & & 0 \end{vmatrix}, \sigma \begin{vmatrix} 1/3 & 1/3 & 1/3 \\ 1/3 & 1/3 & 1/3 \\ 1/3 & 1/3 & 1/3 \end{vmatrix}, \text{ and } \sigma \begin{vmatrix} 1 & 0 & 0 \\ 0 & 0 & 0 \\ 0 & 0 & 0 \end{vmatrix}$$

$$\tag{1.36}$$

for stresses applied along the <110>, <111> and <100> directions, respectively.
   Since perturbation V is a scalar and σ a second-rank tensor, the piezo-
spectroscopic tensor V_{ik} must also be a second-rank tensor.
   Under symmetry operations, the V_{ik} and σ_{ik} have the same transformation
properties. They behave as simple products of coordinates, e.g., V_{xy} transforms as xy, and form a representation of the group. This one can be decomposed into irreducible representations whose basis functions are linear combina-
tions of the V_{ik}. If $V_{\Gamma(i),\alpha}$ and $\sigma_{\Gamma(i),\alpha}$ are, respectively, the αth normalized
basis functions for the ith irreducible representation $\Gamma^{(i)}$, we can write

$$V = \sum_{i,\alpha} V_{\Gamma(i),\alpha}\, \sigma_{\Gamma(i),\alpha} \quad , \tag{1.37}$$

since according to relation (1.34), only products of functions belonging to
the same row of the same representation $\Gamma^{(i)}$ give a nonzero contribution to
V. For instance, in the case of the T_d symmetry for which the irreducible
representations are labelled A_1, A_2, E, T_1, T_2 and the corresponding basis
functions are given in Table 1.2, it can be easily verified that

$$V = V_{A_1}\sigma_{A_1} + V_{E,1}\sigma_{E,1} + V_{E,2}\sigma_{E,2} + 2V_{T_2,1}\sigma_{T_2,1} + 2V_{T_2,2}\sigma_{T_2,2}$$

$$+ 2V_{T_2,3}\sigma_{T_2,3} \quad . \tag{1.38}$$

Indeed, the V_{ik} and σ_{ik} transform like products $x_i x_k$, i.e., V_{xy} behave like
xy, etc. Then,

$$V_{A_1} = \frac{1}{\sqrt{3}} (V_{xx} + V_{yy} + V_{zz}) \quad , \quad \sigma_{A_1} = \frac{1}{\sqrt{3}} (\sigma_{xx} + \sigma_{yy} + \sigma_{zz}) \quad , \tag{1.39}$$

since the basis function for the A_1 representation is $(x^2 + y^2 + z^2)/\sqrt{3}$. In the same way,

$$V_{E,1} = \frac{1}{\sqrt{6}} (2V_{zz} - V_{yy} - V_{xx}) \quad , \quad V_{E,2} = \frac{1}{\sqrt{2}} (V_{xx} - V_{yy}) \quad , \tag{1.40}$$

$$V_{T_2,1} = V_{yz} \ (= V_{zy}) \quad , \quad V_{T_2,2} = V_{zx} \ (= V_{xz}) \quad , \quad V_{T_2,3} = V_{xy} \ (= V_{yx}) \quad ,$$

$$\tag{1.41}$$

and similar expressions hold for σ:

$$\sigma_{E,1} = \frac{1}{\sqrt{6}} (2\sigma_{zz} - \sigma_{yy} - \sigma_{xx}) \quad , \quad \sigma_{E,2} = \frac{1}{\sqrt{2}} (\sigma_{xx} - \sigma_{yy}) \quad , \tag{1.42}$$

$$\sigma_{T_2,1} = \sigma_{yz} \ (= \sigma_{zy}) \quad , \quad \sigma_{T_2,2} = \sigma_{zx} \ (= \sigma_{xz}) \quad , \quad \text{and}$$

$$\sigma_{T_2,3} = \sigma_{xy} \ (= \sigma_{yx}) \quad . \tag{1.43}$$

We want to investigate the first-order splitting due to V on an orbitally degenerate state. The corresponding eigenfunctions are (as we have seen previously) automatically basis functions of an irreducible representation, $\Gamma^{(i)}$. The matrix elements of V are

$$\left\langle \phi_{\Gamma^{(i)},m} \Big| V \Big| \phi_{\Gamma^{(i)},\ell} \right\rangle \quad , \tag{1.44}$$

where $\phi_{\Gamma^{(i)},m}$ is the mth basis function of the orbitally degenerate state corresponding to an irreducible representation $\Gamma^{(i)}$. This matrix can be simplified using the symmetry rules of Sect.1.4.6.

To apply these rules we write this matrix element under the integral form

$$\int \phi^*_{\Gamma^{(i)},m} \phi_{\Gamma^{(j)},\ell} V d\tau \quad , \tag{1.45}$$

which we note

$$\left[\phi^*_{\Gamma^{(i)},m} \phi_{\Gamma^{(j)},\ell} \Big| V \right] \quad , \tag{1.46}$$

where V, being a scalar function, has been commuted with $\phi_{\Gamma^{(j)},\ell}$. We first consider the products $\phi_{\Gamma^{(i)},m} \phi_{\Gamma^{(j)},\ell}$. As we have seen these products form a new representation, which is the direct product $\Gamma^{(i)} \times \Gamma^{(j)}$. This representation can be decomposed into irreducible components $\Gamma^{(k)}$. In the following, we shall label their basis functions $f_{\Gamma(k),m}$. If we now use the expansion (1.37) of V in the stress tensor components we obtain

$$\left\langle \phi_{\Gamma(i),m} \left| V \right| \phi_{\Gamma(i),\ell} \right\rangle = \sum_{j,\alpha} \sigma_{\Gamma(j),\alpha} \left\langle \phi_{\Gamma(i),m} \left| V_{\Gamma(j),\alpha} \right| \phi_{\Gamma(i),\ell} \right\rangle$$

(1.47)

The products $\phi^*_{\Gamma(i),m} \phi_{\Gamma(i),\ell}$ can be expressed as linear combinations of basis functions $f_{\Gamma(k),n}$ and thus each matrix element is a linear combination of terms like

$$\sum_{j,\alpha} \sigma_{\Gamma(j),\alpha} \left[f_{\Gamma(k),n} \left| V_{\Gamma(j),\alpha} \right. \right] \quad , $$

(1.48)

with

$$\left[f_{\Gamma(k),n} \left| V_{\Gamma(j),\alpha} \right. \right] = \int f_{\Gamma(k),n} V_{\Gamma(j),\alpha} \, d\tau \quad . $$

(1.49)

Each such quantity can be considered as the scalar product of $f_{\Gamma(k),n}$ by $V_{\Gamma(j),\alpha}$ and obeys the theorem given in Sect.1.4.6 which states

$$\left[f_{\Gamma(k),n} \left| V_{\Gamma(j),\alpha} \right. \right] = \left[f \middle| V \right]_{\Gamma(k)} \delta_{\Gamma(k),\Gamma(j)} \delta_{n,\alpha} \quad . $$

(1.50)

This allows the reduction of all matrix elements in terms of a minimal number of parameters, the $\left[f \middle| V \right]_{\Gamma(k)}$.

b) Splitting of a Twofold Degenerate E State in the Group T_d

We now consider the tetrahedral point group and the irreducible representation E. There are two basis functions $\phi_{E,1}$ and $\phi_{E,2}$ (Table 1.2). The products $\phi^*_{E,i}\phi_{E,j}$ of these functions belong to the $E \times E$ representation whose characters are $\chi = \chi^2_E$ that is, according to Table 1.2

$$\chi = 4, 1, 4, 0, 0$$

(1.52)

for each different class, respectively. Using (1.18) we then obtain the coefficients

$$a_{A_1} = 1 \quad , \quad a_{A_2} = 1 \quad , \quad a_E = 1 \quad , $$

(1.53)

which means that the representation $E \times E$ decomposes according to

$$E \times E = A_1 + A_2 + E \quad . \tag{1.54}$$

Using the technique given in Sect.1.4.5, it can be easily shown that the basis functions for this representation are

$$f_{A_1} = \frac{1}{2} \left[|\phi_{E,1}|^2 + |\phi_{E,2}|^2 \right] \quad , \tag{1.55}$$

$$f_{E,1} = \frac{1}{2} \left[|\phi_{E,1}|^2 - |\phi_{E,2}|^2 \right] \quad , \tag{1.56}$$

$$f_{E,2} = \frac{1}{2} \left[\phi^*_{E,1}\phi_{E,2} + \phi^*_{E,2}\phi_{E,1} \right] \quad , \quad \text{and} \tag{1.57}$$

$$f_{A_2} = \frac{1}{2} \left[\phi^*_{E,1}\phi_{E,2} - \phi^*_{E,2}\phi_{E,1} \right] \quad . \tag{1.58}$$

We can now calculate the matrix elements of $V_{\Gamma(i),\alpha}$ in the basis of the functions $\phi_{E,1}$ and $\phi_{E,2}$ in order to obtain the matrix

$$\begin{bmatrix} \left[|\phi_{E,1}|^2|V_{\Gamma(i),\alpha} \right] \sigma_{\Gamma(i),\alpha} & , & \left[\phi^*_{E,1}\phi_{E,2}|V_{\Gamma(i),\alpha} \right] \sigma_{\Gamma(i),\alpha} \\ \left[\phi^*_{E,2}\phi_{E,1}|V_{\Gamma(i),\alpha} \right] \sigma_{\Gamma(i),\alpha} & , & \left[|\phi_{E,2}|^2|V_{\Gamma(i),\alpha} \right] \sigma_{\Gamma(i),\alpha} \end{bmatrix} \quad . \tag{1.59}$$

Consider first the case of V_{A_1}. We have the following relations

$$\left[f_{\Gamma(i),m}|V_{A_1} \right] = \left[f_{A_1}|V_{A_1} \right] \delta_{\Gamma(i)_{m,A_1}} \quad , \tag{1.60}$$

with $f_{\Gamma(i),m}$ equal to $f_{A_1}, f_{E_1}, f_{E_2}, f_{A_2}$. The use of the four relations contained in (1.60) allows to write, using (1.55-58), the $f_{\Gamma(i),m}$ in terms of $\phi_{E,1}$ and $\phi_{E,2}$

$$\frac{1}{2} \left[|\phi_{E,1}|^2 + |\phi_{E,2}|^2|V_{A_1} \right] \neq 0 = \left[f_{A_1}|V_{A_1} \right] \quad , \tag{1.61}$$

$$\left[|\phi_{E,1}|^2 - |\phi_{E,2}|^2|V_{A_1} \right] = 0 \quad , \tag{1.62}$$

$$\left[\phi^*_{E,1}\phi_{E,2} + \phi^*_{E,2}\phi_{E,1}|V_{A_1} \right] = 0 \quad , \tag{1.63}$$

$$\left[\phi^*_{E,1}\phi_{E,2} - \phi^*_{E,2}\phi_{E,1}|V_{A_1} \right] = 0 \quad . \tag{1.64}$$

The last two relations imply that only V_{A_1} has diagonal matrix elements. Moreover, the second one shows that the two diagonal terms are equal.

This allows writing the matrix in terms of only one parameter $\left[f_{A_1}|V_{A_1}\right]$. It is, however, interesting to go a little further to obtain the results under the form tabulated in [1.9]. For this we use the expression of the $V_{\Gamma(i),\alpha}$ given above and notice that according to our general theorem, we have

$$\left[f_{A_1}|V_{\Gamma(i),\alpha}\right] = \delta_{A_1,\Gamma(i),\alpha}\left[f_{A_1}|V_{A_1}\right] \quad . \tag{1.65}$$

This allows us to write

$$\left[f_{A_1}\left|\frac{V_{xx}+V_{yy}+V_{zz}}{\sqrt{3}}\right.\right] = \left[f_{A_1}|V_{A_1}\right] \quad , \tag{1.66}$$

$$\left[f_{A_1}|2V_{zz}-V_{xx}-V_{yy}\right] = 0 \quad , \tag{1.67}$$

$$\left[f_{A_1}|V_{xx}-V_{yy}\right] = 0 \quad . \tag{1.68}$$

From these relations we immediately obtain

$$\left[f_{A_1}|V_{A_1}\right] = \sqrt{3}\left[f_{A_1}|V_{xx}\right] \quad . \tag{1.69}$$

If we label A the parameter $\left[f_{A_1}|V_{xx}\right]$, it is immediately apparent that the two diagonal matrix elements of V_{A_1} are equal to $\sqrt{3}$ A. This gives, for the whole perturbation matrix corresponding to $V_{A_1}\sigma_{A_1}$, the final result

$$\begin{bmatrix} A\left(\sigma_{xx}+\sigma_{yy}+\sigma_{zz}\right) & 0 \\ 0 & A\left(\sigma_{xx}+\sigma_{yy}+\sigma_{zz}\right) \end{bmatrix} \quad , \tag{1.70}$$

where we have used the detailed expression (1.39) for σ_{A_1}.

We now consider the matrix elements of $V_{E,1}\sigma_{E,1}$ and $V_{E,2}\sigma_{E,2}$ which correspond to the E irreducible representation. Again we use the fact that

$$\left[f_{E,2}|V_{A_1}\right] = 0 \quad , \tag{1.71}$$

$$\left[f_{E,2}|V_{E,1}\right] = 0 \quad , \tag{1.72}$$

to deduce

$$\left[f_{E,2}|V_{E,2}\right] = \sqrt{6}B \quad , \tag{1.73}$$

where the expression of B is

$$B = \frac{1}{\sqrt{3}} \left\langle \phi_{E,1} | V_{xx} | \phi_{E,2} \right\rangle \quad , \tag{1.74}$$

and the matrix for $V_{E,2}\sigma_{E,2}$ is

$$\begin{bmatrix} 0 & \sqrt{6}B\sigma_{E,2} \\ \sqrt{6}B\sigma_{E,2} & 0 \end{bmatrix} . \tag{1.75}$$

In the same way,

$$\left[f_{E,1} | V_{E,2} \right] = \left[f_{E,2} | V_{E,2} \right] = \sqrt{6}B \quad \text{and} \tag{1.76}$$

$$\left[f_{E,2} | V_{E,1} \right] = 0 \tag{1.77}$$

lead to the following matrix for the contribution of $V_{E,1}\sigma_{E,1}$:

$$\begin{bmatrix} -\sqrt{6}B\sigma_{E,1} & 0 \\ 0 & \sqrt{6}B\sigma_{E,1} \end{bmatrix} . \tag{1.78}$$

Finally, the matrix elements of the terms $V_{T_2,i}\sigma_{T_2,i}$ which belong to the T_2 irreducible representation, all vanish. This results from the fact that the products $\phi^*_{E,i}\phi_{E,j}$, as we have seen, form a basis of a representation $E \times E$ which decomposes into $A_1 + A_2 + E$ and therefore does not contain T_2. There are no $f_{\Gamma(i),m}$ with $\Gamma^{(i)}$ equal to T_2 and thus no term like $\left[f_{T_2,m} | V_{T_2,m} \right]$ which would be nonzero by symmetry.

All the results corresponding to the splitting of the E state under stress are summarized in Table 1.6. The matrices corresponding to <100>, <110>, and <111> applied stress are given. The notations are the same as in [1.9].

c) The case of a T_2 State

A similar procedure should be used for the basis functions belonging to the T_2 representation. The characters of the $T_2 \times T_2$ representation (9,0,1,1,1) give the following decomposition

$$T_2 \times T_2 = A_1 + E + T_1 + T_2 \quad . \tag{1.79}$$

The corresponding basis functions built from the basis functions ϕ_x, ϕ_y and ϕ_z of the T_2 representation are given in Table 1.7. The third parameter C introduced in this table is

Table 1.6. Matrices $V_\Gamma(i)_{,\alpha}\ \sigma_\Gamma(i)_{,\alpha}$ for an E state

The first row gives the complete matrices, the other rows their expressions for <100>, <111>, and <110> stress. The last two columns give the corresponding eigenvalues and degeneracy

	$V_{A_1}\ \sigma_{A_1}$		$V_{E_1}\ \sigma_{E_1}$		$V_{E_2}\ \sigma_{E_2}$		Eigenvalues	Degeneracy
	$A(\sigma_{xx}+\sigma_{yy}+\sigma_{zz})$	0	$-B(2\sigma_{zz}-\sigma_{xx}-\sigma_{yy})$	0	0	$\sqrt{3}\,B(\sigma_{xx}-\sigma_{yy})$		
	0	$A(\sigma_{xx}+\sigma_{yy}+\sigma_{zz})$	0	$+B(2\sigma_{zz}-\sigma_{xx}-\sigma_{yy})$	$\sqrt{3}\,B(\sigma_{xx}-\sigma_{yy})$	0		
$\sigma\parallel$ <100>	A	0	B	0	0	$\sqrt{3}\,B$	$A - 2B$	1
	0	A	0	$-B$	$\sqrt{3}\,B$	0	$A + 2B$	1
$\sigma\parallel$ <111>	A	0	0	0	0	0	A	2
	0	A	0	0	0	0		
$\sigma\parallel$ <110>	A	0	B	0	0	0	$A + B$	1
	0	A	0	$-B$	0	0	$A - B$	1

Basis functions

$$f_{A_1} = \frac{|\varphi_{E,1}|^2 + |\varphi_{E,2}|^2}{2}$$

$$f_{E,1} = \frac{|\varphi_{E,1}|^2 - |\varphi_{E,2}|^2}{2}$$

$$f_{E,2} = \frac{\varphi_{E,1}^*\varphi_{E,2} + \varphi_{E,2}^*\varphi_{E,1}}{2}$$

$$f_{A_2} = \frac{\varphi_{E,1}^*\varphi_{E,2} - \varphi_{E,2}^*\varphi_{E,1}}{2}$$

Table 1.7. Matrices $V_{\Gamma_i,\alpha}$ $\sigma_{\Gamma_i,\alpha}$ for a T_2 state. (The notations are the same as in Table 1.6)

Header matrices:

$V_{A_1}\,\sigma_{A_1}$:
$$\begin{pmatrix} A(\sigma_{xx}+\sigma_{yy}+\sigma_{zz}) & 0 & 0 \\ 0 & A(\sigma_{xx}+\sigma_{yy}+\sigma_{zz}) & 0 \\ 0 & 0 & A(\sigma_{xx}+\sigma_{yy}+\sigma_{zz}) \end{pmatrix}$$

$V_{E_1}\,\sigma_{E_1}$:
$$\begin{pmatrix} \frac{B}{2}(2\sigma_{zz}-\sigma_{xx}-\sigma_{yy}) & 0 & 0 \\ 0 & -\frac{B}{2}(2\sigma_{zz}-\sigma_{xx}-\sigma_{yy}) & 0 \\ 0 & 0 & B(2\sigma_{zz}-\sigma_{xx}-\sigma_{yy}) \end{pmatrix}$$

$V_{E_2}\,\sigma_{E_2}$:
$$\begin{pmatrix} 3/2\,B(\sigma_{xx}-\sigma_{yy}) & 0 & 0 \\ 0 & -3/2\,B(\sigma_{xx}-\sigma_{yy}) & 0 \\ 0 & 0 & 0 \end{pmatrix}$$

$\sum_\alpha V_{T_2,\alpha}\,\sigma_{T_2,\alpha}$:
$$\begin{pmatrix} 0 & 2C\,\sigma_{xy} & 2C\,\sigma_{xz} \\ 2C\,\sigma_{xy} & 0 & 2C\,\sigma_{yz} \\ 2C\,\sigma_{xz} & 2C\,\sigma_{yz} & 0 \end{pmatrix}$$

	$V_{A_1}\,\sigma_{A_1}$	$V_{E_1}\,\sigma_{E_1}$	$V_{E_2}\,\sigma_{E_2}$	$\sum_\alpha V_{T_2,\alpha}\,\sigma_{T_2,\alpha}$	Eigenvalues	Degeneracy
$\sigma\parallel\langle100\rangle$	$\begin{pmatrix} A&0&0\\0&A&0\\0&0&A \end{pmatrix}$	$\begin{pmatrix} B/2&0&0\\0&B/2&0\\0&0&-B \end{pmatrix}$	$\begin{pmatrix} 3B/2&0&0\\0&-3B/2&0\\0&0&0 \end{pmatrix}$	$\begin{pmatrix} 0&0&0\\0&0&0\\0&0&0 \end{pmatrix}$	$A+2B$ $A-B$	1 2
$\sigma\parallel\langle111\rangle$	$\begin{pmatrix} A&0&0\\0&A&0\\0&0&A \end{pmatrix}$	$\begin{pmatrix} 0&0&0\\0&0&0\\0&0&0 \end{pmatrix}$	$\begin{pmatrix} 0&0&0\\0&0&0\\0&0&0 \end{pmatrix}$	$\begin{pmatrix} 0&2C/3&2C/3\\2C/3&0&2C/3\\2C/3&2C/3&0 \end{pmatrix}$	$A+2C/3$ $A-C/3$	1 2
$\sigma\parallel\langle110\rangle$	$\begin{pmatrix} A&0&0\\0&A&0\\0&0&A \end{pmatrix}$	$\begin{pmatrix} B/2&0&0\\0&B/2&0\\0&0&-B \end{pmatrix}$	$\begin{pmatrix} 0&0&0\\0&0&0\\0&0&0 \end{pmatrix}$	$\begin{pmatrix} 0&C&0\\C&0&0\\0&0&0 \end{pmatrix}$	$A+(B+C)/2$ $A+(B-C)/2$ $A-B$	1 1 1

Basis functions:

$$f_{A_1} = \frac{1}{\sqrt{3}}\left(|\varphi_x|^2 + |\varphi_y|^2 + |\varphi_z|^2\right)$$

$$f_{E,1} = \frac{1}{\sqrt{6}}\left(2|\varphi_z|^2 - |\varphi_x|^2 - |\varphi_y|^2\right)$$

$$f_{E,2} = \frac{1}{\sqrt{2}}\left(|\varphi_x|^2 - |\varphi_y|^2\right)$$

$$f_{T_2,x} = \frac{1}{2}\left(\varphi_y^*\varphi_z + \varphi_z^*\varphi_y\right)$$

$$f_{T_2,y} = \frac{1}{2}\left(\varphi_x^*\varphi_z + \varphi_z^*\varphi_x\right)$$

$$f_{T_2,z} = \frac{1}{2}\left(\varphi_x^*\varphi_y + \varphi_y^*\varphi_x\right)$$

$$f_{T_1,1} = \frac{1}{2}\left(\varphi_y^*\varphi_z - \varphi_z^*\varphi_y\right)$$

$$f_{T_1,2} = \frac{1}{2}\left(\varphi_x^*\varphi_z - \varphi_z^*\varphi_x\right)$$

$$f_{T_1,3} = \frac{1}{2}\left(\varphi_x^*\varphi_y - \varphi_y^*\varphi_x\right)$$

$$C = <\phi_y|V_{T_2,x}|\phi_z> = <\phi_z|V_{T_2,y}|\phi_x> = <\phi_x|V_{T_2,z}|\phi_y> \quad . \tag{1.80}$$

The A and B parameters, introduced previously, are still given by relation (1.69) and (1.73), but with the f_{A_1} and f_{E2} defined in Table 1.7.

The eigenvalues of the different matrices obtained when stresses are applied in the three <100>, <110>, and <111> directions are given in Tables 1.6,7 for E and T_2 orbitally degenerate states. The results are summarized in Fig.1.16, showing the way the splitting of the degenerate levels occurs.

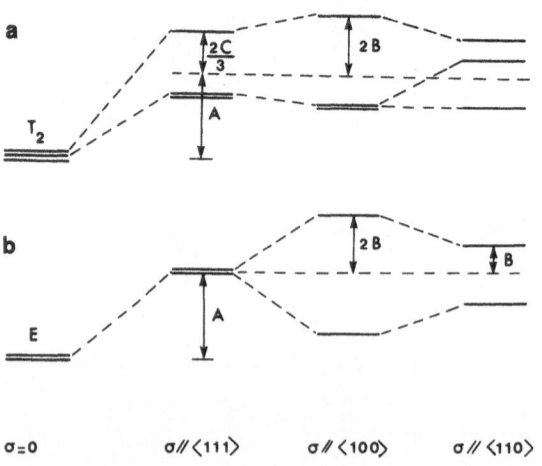

<u>Fig.1.16a,b.</u> Splitting of degenerate (a) T_2 and (b) E states under uniaxial stress applied along the three main crystallographic axes. The notations are defined in Tables 1.6,7

d) Lowering of Symmetry and Orientational Degeneracy

Many defects have a symmetry lower than the symmetry of their site in the perfect crystal. As we have seen, this can result from a spontaneous distortion leading to a more stable situation. This also occurs when there is association of defects, for instance, vacancy-substitutional impurity pairs. The symmetry group is then a subgroup of the original one. For instance, there are many cases in the diamond lattice where the symmetry group T_d becomes its subgroup D_{2d} [tetragonal, along (100) directions] or C_{3v} [trigonal, along (111) directions].

Such a lowering of symmetry has important consequences as regards the stress-splitting experiments. First, the degeneracy of the levels is lower since the dimensionality of the irreducible representations is lower. There

will be less splitting for one given defect with one particular orientation.
However, there are several equivalent orientations (three for D_{2d}, four for
C_{3v}). They will respond differently to a given applied stress. This will lead
to another kind of splitting, resulting from what is called orientational degeneracy.

To illustrate the effect of orientational degeneracy, let us consider a
nondegenerate level for a tetragonal center for which there are three orientations along the <100> directions. These orientations remain equivalent under
a stress applied along the <111> direction. This is no longer true for a <100>
stress, in which case the <100> level will shift differently from the <010>
and <001> one. This is also true for a <110> stress. Figure 1.17 summarizes
this effect for <100> and <111> defects and shows the relative weights of the
split levels. Clearly, such splittings are characteristic of the defect symmetry.

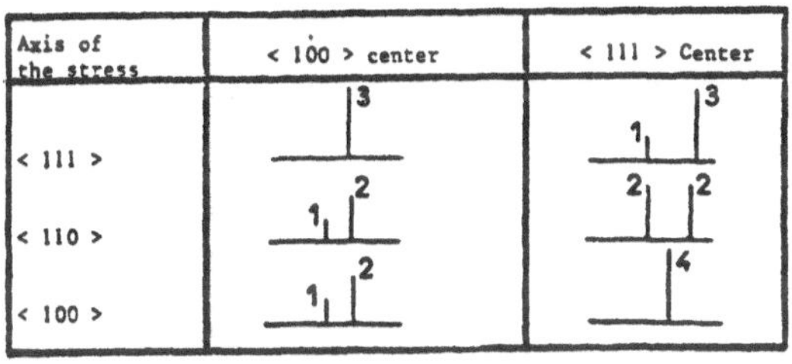

Axis of the stress	< 100 > center	< 111 > Center
< 111 >		
< 110 >		
< 100 >		

Fig.1.17. Splitting of levels associated with <100> and <111> oriented defects. The numbers indicate the relative amplitudes of the splitted lines

The case of level plus orientational degeneracy is more complex. It is
necessary to consider each crystallographic direction separately and to write
down the perturbation matrices associated with the stress. For this we proceed exactly as we have done for the group T_d, but with the irreducible representations and basis functions of the given subgroup. Identical results can
be obtained in a simpler way by using the results obtained for T_d. For this
we apply a fictitious stress σ_0 whose effect is to lower the symmetry to the
desired subgroup. For instance, if σ_0 is along a <100> axis, the resulting
subgroup is D_{2d} and the other two equivalent situations correspond to σ_0
along <010> and <001>. The resulting splitting is given by Tables 1.6,7. It
is the same for any equivalent crystallographic direction of the defect, but
with different basis functions. To this zero stress splitting, we now have
to add the extrasplitting due to the applied stress σ. The final observed
splitting then must correspond to a total stress $\sigma_0 + \sigma$ with varying equivalent

σ_0 to simulate the equivalent crystallographic directions. A simple illustration of this is given on Fig.1.18 for a T_2 state of the group T_d.

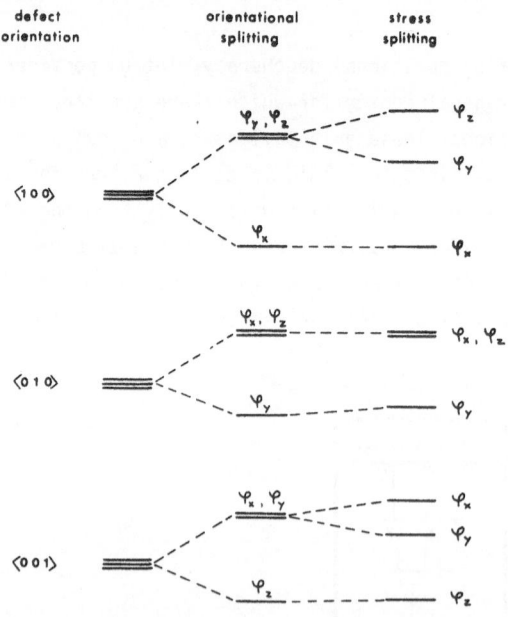

Fig.1.18. Effect of a <010> stress on the T_2 level of a defect whose electric dipole is oriented in three <100> directions

1.5.2 Dipolar Transitions

The intensity of an optical transition between an initial state ψ_i and a final state ψ_j is proportional the square of the matrix element

$$< \psi_i |D| \psi_j > \quad , \tag{1.81}$$

where D is the dipole moment of the system. For one-electron states ψ_i and ψ_j, the intensity is simply the product $\underline{n} \cdot \underline{r}$, where $\underline{n}$ is the unit vector of the direction of polarization and $\underline{r}$ the position vector of the electron. In this case, the calculation of this intensity requires knowledge of the matrix elements of x,y,z, the components of $\underline{r}$ which belong to the representation T_2, between the basis functions of the levels ψ_i and ψ_j. The procedure for calculating these elements is completely analogous to the calculation of the matrix elements of the perturbation produced by a stress $\sigma_{T_2,i}$.

Consider the case of the T_d group developed in the preceeding section. For two basis functions of the representation E, $\psi_i^* \psi_j$ belongs to $E \times E$ which does not contain T_2, and consequently, there are no transitions between E levels. Since $T_2 \times T_2$ contains the T_2 representation, the matrix $<\psi_i|D|\psi_j>$ for the transition between the $\psi_{T_2,x}$, $\psi_{T_2,y}$, $\psi_{T_2,z}$ states is obviously

$$\begin{bmatrix} 0 & qn_z & qn_y \\ qn_z & 0 & qn_x \\ qn_y & qn_x & 0 \end{bmatrix} \quad , \tag{1.82}$$

where

$$q = <\psi_{T_2,y}|x|\psi_{T_2,z}> \quad . \tag{1.83}$$

It is clear that the states between which transitions can occur depend upon the direction of polarization. Splitting of the T_2 level by distortion and applied stress will not alter these selection rules, which therefore, apply also to the cases described above.

1.5.3 Other Excitations

In the case of application of an electric field $\underline{E}$ (Stark effect) the perturbation is described by

$$\delta H = e \cdot \underline{r} \cdot \underline{E} \quad , \tag{1.84}$$

where e is the electronic charge, and $\underline{r}$ the displacement vector. The quantity $<\psi|H|\psi>$ is then nonzero only for the defects that do not possess a center of symmetry. In that case, the shift or the decomposition of optical transitions which are induced is usually small since even for high electric field (typically 10^5 V cm^{-1}) the quantity δH remains small, often smaller than the linewidth of the transition.

The application of a magnetic field (Zeeman effect), which is limited to the case of paramagnetic centers, will not be described here since it is fully developed in Vol.2, Chap.9. We shall see there that electron paramagnetic resonance provides the A (hyperfine) and D (spin) tensors which give directly the mapping of the wave function associated with the defect and its character (s- or p-like for instance). For a more complete discussion see [1.10].

2. Effective Mass Theory

If we consider an isolated defect in an otherwise perfect crystal, the resulting effects on the eigenvalues of the electron Schrödinger equation are
the following.

a) The defect can introduce new energy levels in one of the forbidden
   energy gaps of the perfect crystal. These are bound or localized states
   whose wave function is exponentially decreasing away from the defect
   site.

b) The defect can scatter the electrons whose energies lie within the
   allowed energy bands of the crystal. The resulting effect can be described as a change in density of states due to the presence of the
   defect.

The most striking effect is the appearance of bound states. At the moment,
there exists no general theory for dealing with this problem. However, there
are different methods of approximation which apply in different limiting cases.
The first of them, the effective mass theory, is well-adapted to the study of
shallow levels characteristic of the donor and acceptor impurities. Its great
advantage is its simplicity, since the bound-state equation reduces to an
hydrogenic Schrödinger equation. The other theories are adapted to more strongly bound states lying deeper in the gap. In general, the simplification results
from the strong localization of the potential. In this chapter we restrict
ourselves to the effective mass theory, leaving the deep-level treatments for
Chaps. 3 and 4.

We begin by discussing the effective mass theory in a simplified manner
for the case of donors. We will then present a justification for a one-band
system and its extension to the real situation of donors in group IV elements, where there exist several equivalent minima of the conduction band.
We will also give a short discussion on acceptor states. The results are then

used to analyze the experimental data. Finally, we will give some details
about the case of electrons or electron-hole pairs bound to donor-acceptor
pairs in semiconductors. The entire derivation will be presented in real
space since we believe that it gives a better physical insight into the nature of the approximation. This procedure has already been used for the one-
band case [2.1,3], and we will generalize it to more complex situations.
Reciprocal space derivations can be found in many textbooks [2.4,6] or review articles [2.7,8].

2.1 Simplified Presentation

We discuss here the case of donors in Group IV covalent semiconductors. These
semiconductors are characterized by very strong covalent bonds. They are
schematically represented in Fig.2.1. Each atom has four equivalent bonds with
its nearest neighbors because it possesses four valence electrons engaging
one electron per bond. Each covalent bond thus contains two electrons. This

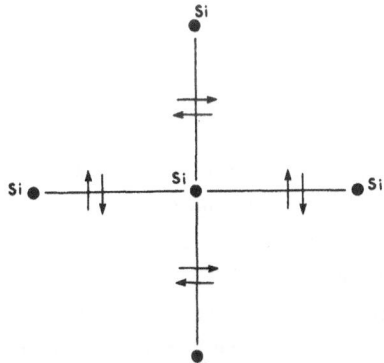

Fig.2.1. Schematic representation of the
localization of valence electrons in a
covalent material

is completely equivalent to the concept of covalency developed for molecules.
(For general treatments of the covalent bond see [2.9,11].) Such a description is in fact quite correct and it can be shown (Chap.3) that in each bond,
the two electrons are in the same bonding orbital with opposite spins [2.12].
This view has been confirmed by sophisticated calculations [2.13]. An important feature of this covalent bonding is that the gain in energy is important.
The energy per bond is discussed in Chap.6 and found to be 7.9 eV for diamond,
5.9 eV for silicon and 6 eV for germanium. This means that very strong perturbations are needed to break such bonds.

   Let us now introduce a substitutional donor in the lattice, i.e., replace
the central atom of Fig.2.1 by an atom of Group V (phosphorus for instance).
This atom has one excess nuclear charge +e relative to a Group IV atom and
one excess valence electron (this is also true for other Group V impurities
as long as core electrons are neglected). In view of the great stability of
the covalent bond, the Group V atom will still form four such bonds with its
neighbors. This means that four of its electrons will be engaged in bonding
states, i.e., will have energies falling within the valence band. We are thus
left with an extra electron that should be in a conduction-band state, unless
new energy levels appear in the forbidden gap. To investigate such a possibility, we have to write the Schrödinger equation for the excess electron notic-
ing that the main change seen by this electron corresponds to the potential
of the excess nuclear charge +e on the impurity. In a vacuum, this is exactly
the situation of the hydrogen atom. In a solid, however, many corrections must
be made to that simple picture. They are the following.

   a) Because of the dielectric constant of the medium, the nuclear charge
      +e polarizes the solid, leading to dielectric screening. The potential
      experienced by the excess electron is not $-e^2/r$ but $-e^2/\varepsilon r$, where ε is
      the static macroscopic dielectric constant.
   b) Usually in the hydrogen atom, the origin of the energies is taken at
      infinite separation. The ground-state energy is just the opposite of
      the ionization energy necessary for the electron to become free. Here
      the lowest energy at which the electron becomes free is the bottom of
      the conduction band E_c, which is thus the required origin of energies.
   c) Once the electron is ionized in a vacuum, it propagates with its own
      mass m. This is not true in the solid where the periodic potential in-
      fluences this propagation. This can be taken into account by introducing
      the concept of the "effective mass" m^*, which generally is direction
      dependent (i.e., m^* is a second rank tensor). For the time being, we
      shall ignore this anisotropy of m^*.

   Incorporating these three fundamental corrections leads for the excess
electron to the following hydrogenic equation

$$\left(\frac{\hbar^2}{2m^*}\Delta - \frac{e^2}{\varepsilon r}\right)\psi = (E - E_c)\psi \quad . \tag{2.1}$$

$\hbar = h/2\pi$ (normalized Planck's constant)

The solutions of this equation are obtained from the hydrogen case by replacing m by m^* and e^2 by e^2/ε. The energy levels are thus given by the values of the hydrogen atom, scaled for m^* and ε that is,

$$E_n = E_c - 13.6 \, \frac{m^*}{m\varepsilon^2} \cdot \frac{1}{n^2} \, [\text{eV}] \quad , \tag{2.2}$$

where n is an integer. A typical case corresponds to $m^* \approx 0.1$ m, $\varepsilon \approx 10$, which leads to a ground state lying 13.6 meV under the conduction band, a very small quantity compared to the forbidden band gap in silicon and germanium (~1 eV). We can also derive the ground-state wave function, which is given by the hydrogen 1s ground state

$$\psi_{1s}(r) = A \, e^{-r/a} \tag{2.3}$$

where the Bohr radius a is the scaled one, i.e.,

$$a = 0.53 \, \frac{\varepsilon}{m^*/m} \, [\text{Å}] \quad . \tag{2.4}$$

In our typical case, this leads to a Bohr radius of 53 Å, a value quite large when compared to the interatomic distance (~2.3 Å).

2.2 Derivation in the One-Band Case

2.2.1 The Case of One Band with One Extremum

For simplicity we start from a perfect crystal with one atom per unit cell and assume there is only one energy band $E(\underline{k})$ of interest. The energy values $E(\underline{k})$ are solutions of the Schrödinger equation:

$$H_0 \psi_{\underline{k}} = E(\underline{k}) \psi_{\underline{k}} \quad , \tag{2.5}$$

where H_0 is the perfect crystal hamiltonian. The wave function $\psi_{\underline{k}}$ represents a propagating Bloch state of wave vector $\underline{k}$. We shall express the Schrödinger equation for the impurity in real space, instead of the reciprocal as usually done. For this, it is necessary to define Wannier functions $W(\underline{r} - \underline{R}_i)$, the Fourier transforms of the Bloch states $\psi_{\underline{k}}(\underline{r})$:

$$W(\underline{r} - \underline{R}_i) = \frac{1}{\sqrt{N}} \sum_{\underline{k}} \exp\left[-i\underline{k} \cdot \underline{R}_i\right] \psi_{\underline{k}}(\underline{r}) \quad . \tag{2.6}$$

Here N is the number of atoms and R_i are the lattice vectors. Equation (2.6) can be inverted to give

$$\psi_{\underline{k}}(\underline{r}) = \frac{1}{\sqrt{N}} \sum_{\underline{R}_j} \exp\left[i\underline{k} \cdot \underline{R}_j\right] W(\underline{r} - \underline{R}_j) \quad . \tag{2.7}$$

The Wannier functions defined in this way are obtained by a unitary transformation from $\psi_{\underline{k}}(\underline{r})$. Thus, they form an orthonormal set. Being Fourier transforms of delocalized wave functions, they have the important advantage of being localized in real space. We notice that (2.7) is the expression found for the wave function in the linear combination of atomic orbitals (LCAO) method with one basis function per atom. The only difference is that the Wannier functions are orthonormal, while the atomic functions are not (however, it is always possible to orthogonalize them by the method of LÖWDIN [2.14]). We can express $E(\underline{k})$ in terms of the matrix elements of H_0 in the Wannier basis by projecting (2.5) onto $W(\underline{r} - \underline{R}_0)$. This leads to

$$E(\underline{k}) = \sum_{\underline{R}_h} \langle W(\underline{r} - \underline{R}_0) | H_0 | W(\underline{r} - \underline{R}_h) \rangle \exp(i\underline{k} \cdot \underline{R}_h) \quad , \tag{2.8}$$

that is

$$E(\underline{k}) = \sum_{\underline{R}_h} H_0(\underline{R}_0, \underline{R}_h) \exp(i\underline{k} \cdot \underline{R}_h) \quad . \tag{2.9}$$

In this section we restrict ourselves to the case of attractive potentials for which the localized energy level at energy E falls below the allowed energy band. This band is pictured schematically in Fig.2.2 versus the vector $\underline{k}$ in a direction containing the absolute minimum E_c which occurs at a value of $\underline{k}$ equal to $\underline{k}_0$.

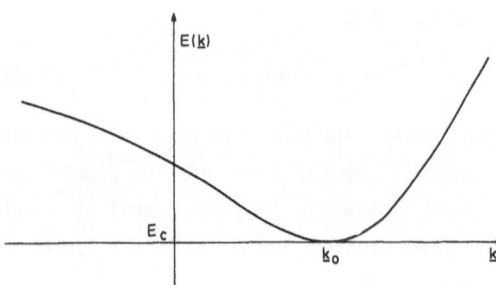

Fig.2.2. Schematic representation of band energy versus wave vector

The Schrödinger equation corresponding to the crystal containing the impurity is

$$(H_0 + U)\psi = E\psi \quad , \tag{2.10}$$

where $U(\underline{r})$ is the perturbation introduced by the defect. The first step in solving this equation consists in expressing $\psi(\underline{r})$ in the basis formed by the perfect-crystal wave functions. We use the Wannier functions $W(\underline{r} - \underline{R}_i)$ and write

$$\psi(\underline{r}) = \sum_{\underline{R}_i} c(\underline{R}_i)W(\underline{r} - \underline{R}_i) \quad , \tag{2.11}$$

$c(\underline{R}_i)$ being unknown coefficients. Before going on, a very important point must be made here. In the limit where U becomes extremely small and slowly varying, the localized state energy E will tend towards $E(\underline{k}_0)$ and the corresponding wave function towards ψ_{k_0}. In this limit, the coefficients $c(\underline{R}_i)$ become proportional to $\exp(i\underline{k}_0 \cdot \underline{R}_i)$, the coefficients of the expansion of the Wannier functions. We can thus write

$$c(R_i) = \exp(i\underline{k}_0 \cdot \underline{R}_i)f(\underline{R}_i) \quad , \tag{2.12}$$

and ψ becomes

$$\psi(\underline{r}) = \sum_{\underline{R}_i} \exp(i\underline{k}_0 \cdot \underline{R}_i)f(\underline{R}_i)W(\underline{r} - \underline{R}_i) \tag{2.13}$$

The important property of $f(\underline{R}_i)$ we shall use in the following is that, in the same limit (U small), it tends towards a constant.

We can now introduce the expansion (2.13) of the wave function in the Schrödinger equation (2.10). This leads to

$$\sum_{\underline{R}_j} \left\langle W(\underline{r} - \underline{R}_i)|H_0 + U(\underline{r})|W(\underline{r} - \underline{R}_j)\right\rangle \exp(i\underline{k}_0 \cdot \underline{R}_j)f(\underline{R}_j)$$

$$= E \exp(i\underline{k}_0 \cdot \underline{R}_i)f(\underline{R}_i) \quad . \tag{2.14}$$

In this expression, we write $\underline{R}_j$ as $\underline{R}_i + \underline{R}_h$ and obtain

$$\sum_{\underline{R}_h} \left\langle W(\underline{r} - \underline{R}_i)|H_0 + U(\underline{r})|W(\underline{r} - \underline{R}_i - \underline{R}_h)\right\rangle \exp(i\underline{k}_0 \cdot \underline{R}_h)f(\underline{R}_i + \underline{R}_h)$$

$$= E f(\underline{R}_i) \tag{2.15}$$

or equivalently,

$$\sum_{\underline{R}_h} \left[H_0(\underline{R}_0,\underline{R}_h) + U(\underline{R}_i,\underline{R}_i + \underline{R}_h)\right] \exp(i\underline{k}_0 \cdot \underline{R}_h)f(\underline{R}_i + \underline{R}_h) = E f(\underline{R}_i) \quad ,$$

$$\tag{2.16}$$

where $U_0(\underline{R}_i,\underline{R}_i+\underline{R}_h)$ is the matrix element of $U(\underline{r})$ between the two Wannier functions centered at sites $\underline{R}_i$ and $\underline{R}_i+\underline{R}_h$. For H_0 we use the same notation, but the translational symmetry reduces this matrix element to $H_0(\underline{R}_0,\underline{R}_0+\underline{R}_h)$.

In the limit of a small and slowly varying perturbation $U(\underline{r})$, the unknown coefficients $f(\underline{R}_i)$ tend towards a constant and thus vary very little with R_i. In this limit, it is possible to replace the discrete set of equations by a continuous equation. We consider that $f(\underline{R}_i)$ is the value that the function $f(\underline{x})$ of the continuously variable vector $\underline{x}$ will take at $\underline{x} = \underline{R}_i$. Equation (2.16) is then equivalent to the following continuous equation:

$$\sum_{\underline{R}_h} \left[H_0(\underline{R}_0,\underline{R}_h) + U(\underline{x},\underline{x}+\underline{R}_h) \right] \exp(i\underline{k}_0 \cdot \underline{R}_h) f(\underline{x}+\underline{R}_h) = E\, f(\underline{x}) \quad , \qquad (2.17)$$

whose solution taken at the points $\underline{x} = \underline{R}_i$ gives the unknown coefficients $f(\underline{R}_i)$. To go further, we could transform again this equation (which is still exact) to obtain a more compact but formal expression. Here we prefer to introduce directly the two decisive approximations, leading to the usual version of effective mass theory for substitutional impurities.

The fundamental approximation concerns the first term on the left in (2.17). In this term $H_0(\underline{R}_0,\underline{R}_h)$ decreases rapidly with $\underline{R}_h$ since the Wannier functions are localized, and $f(\underline{x}+\underline{R}_h)$ is expected to be a slowly varying function as discussed above. We can therefore expand $f(\underline{x}+\underline{R}_h)$ in powers series of $\underline{R}_h$. This leads to

$$\sum_{\underline{R}_h} H_0(\underline{R}_0,\underline{R}_h) \exp(i\underline{k}_0 \cdot \underline{R}_h) f(\underline{x}+\underline{R}_h) = \sum_{\underline{R}_h} H_0(\underline{R}_0,\underline{R}_h) \exp(i\underline{k}_0 \cdot \underline{R}_h)$$

$$\left[1 + \sum_\alpha X_h^\alpha \frac{\partial}{\partial x^\alpha} f(x) + 1/2 \sum_{\alpha\beta} X_h^\alpha X_h^\beta \frac{\partial^2}{\partial x^\alpha \partial x^\beta} f(x)... \right] \quad , \qquad (2.18)$$

where X_h^α and x^α are the αth components of $\underline{R}_h$ and $\underline{x}$, respectively.

From the comparison of (2.18) with the expression $E(\underline{k})$ given by (2.9), it is easy to see that

$$\sum_{\underline{R}_h} H_0(\underline{R}_0,\underline{R}_h) \exp(i\underline{k}_0 \cdot \underline{R}_h) = E(\underline{k}_0) \quad , \qquad (2.19)$$

$$\sum_{\underline{R}_h} H_0(\underline{R}_0,\underline{R}_h) \exp(i\underline{k}_0 \cdot \underline{R}_h) X_h^\alpha = -i\left(\frac{\partial E(\underline{k})}{\partial k_\alpha} \right)_{\underline{k}=\underline{k}_0} \quad , \qquad (2.20)$$

and

$$\sum_{\underline{R}_h} H_0(\underline{R}_0, \underline{R}_h) \, \exp(i\underline{k}_0 \cdot \underline{R}_h) x_h^\alpha x_h^\beta = -\left(\frac{\partial^2 E(\underline{k})}{\partial k_\alpha \partial k_\beta}\right)_{\underline{k}=\underline{k}_0} \qquad (2.21)$$

Since $\underline{k}_0$ corresponds to the minimum of $E(\underline{k})$, the first-order derivative given by (2.20) vanishes at this point. Let us define the effective mass tensor M by its inverse M^{-1}, that is,

$$\hbar^2 M_{\alpha\beta}^{-1} = \left(\frac{\partial^2 E}{\partial k_\alpha \partial k_\beta}\right)_{\underline{k}=\underline{k}_0} \qquad (2.22)$$

With this notation, (2.17) becomes

$$-\frac{\hbar^2}{2} \sum_{\alpha\beta} M_{\alpha\beta}^{-1} \frac{\partial^2 f(\underline{x})}{\partial x^\alpha \partial x^\beta} + \sum_{\underline{R}_h} U(\underline{x}, \underline{x} + \underline{R}_h) \exp(i\underline{k}_0 \cdot \underline{R}_h) f(\underline{x} + \underline{R}_h)$$

$$= [E - E(\underline{k}_0)] f(\underline{x}) \qquad (2.23)$$

$M_{\alpha\beta}^{-1}$ is a symmetrical tensor and as such can be diagonalized by a suitable choice of basis vectors. Within this set $M_{\alpha\beta}^{-1}$ has in general three different elements which are the inverses of the effective masses. Equation (2.23) can therefore be written

$$-\frac{\hbar^2}{2} \left(\frac{1}{m_1^*} \frac{d^2}{dx_1^2} + \frac{1}{m_2^*} \frac{d^2}{dx_2^2} + \frac{1}{m_3^*} \frac{d^2}{dx_3^2}\right) f(\underline{x}) + \sum_{\underline{R}_h} U(\underline{x}, \underline{x} + \underline{R}_h)$$

$$\exp(i\underline{k}_0 \cdot \underline{R}_h) f(\underline{x} + \underline{R}_h) = [E - E(\underline{k}_0)] f(\underline{x}) \qquad (2.24)$$

This second-order differential equation is, however, not easily solved if the potential term is not simplified further. This leads to the second approximation (not completely distinct from the first one) which assumes that $U(\underline{r})$ varies slowly with $\underline{r}$ so that its matrix elements $U(\underline{R}_i, \underline{R}_i + \underline{R}_h)$ become equal to

$$U(\underline{R}_i, \underline{R}_i + \underline{R}_h) \simeq U(\underline{R}_i, \underline{R}_i) \delta_{\underline{R}_i, \underline{R}_h} \qquad (2.25)$$

This approximation becomes exact in the limit of infinitely slow variations of the potential $U(\underline{r})$. It has the enormous advantage of eliminating all terms for which $\underline{R}_h \neq 0$, leading to

$$\left[-\frac{\hbar^2}{2}\left(\frac{1}{m_1^*}\frac{d^2}{dx_1^2} + \frac{1}{m_2^*}\frac{d^2}{dx_2^2} + \frac{1}{m_3^*}\frac{d^2}{dx_3^2}\right) + U(\underline{x},\underline{x})\right] f(\underline{x}) = [E - E(\underline{k}_0)] f(\underline{x}) \qquad .$$

$$(2.26)$$

This is the usual form of the effective mass equation. For isotropic systems ($m_1^* = m_2^* = m_3^* = m^*$) it simplifies further, reducing to an ordinary Schrödinger equation

$$\left[-\frac{\hbar^2}{2m^*} \Delta + U(\underline{x},\underline{x}) \right] f(\underline{x}) = [E - E(\underline{k}_0)]f(\underline{x}) \quad . \tag{2.27}$$

The application of this equation to the simple donor of Sect.2.1 for which $U(\underline{r})$ is equal to $-e^2/\varepsilon r$ leads to

$$U(\underline{x},\underline{x}) = -\frac{e^2}{\varepsilon|\underline{x}|} \quad , \tag{2.28}$$

and we recover the hydrogenic equation derived in Sect.2.2 from elementary considerations.

2.2.2 The Case of Equivalent Extrema

The situation we just described does not occur in practice. For instance, the conduction band of the Group IV elements presents several equivalent extrema (six for silicon, eight for germanium). There are also two atoms per unit cell instead of one. It is still possible to define Wannier functions for the lowest conduction band as in (2.6), but these functions will now be Wannier functions for the unit cell and not for a particular atom. The vectors $\underline{R}_i$ will also refer to a unit cell.

The wave function $\psi(\underline{r})$ for the bound state is again written in terms of the Wannier functions for this band as in (2.13). However, in the limit of slowly varying $U(\underline{r})$, we no longer can say the coefficients $c(\underline{R}_i)$ tend towards the value $\exp(i\underline{k}_0 \cdot \underline{R}_i)$ corresponding to the Bloch function at the minimum $\underline{k}_0$ of the band. There are now g equivalent minima corresponding to $\underline{k}_{0m}$, with $m = 1,\ldots,g$. The function $\psi(\underline{r})$ tends towards a combination of the g Bloch functions $\psi_{\underline{k}_{0m}}(\underline{r})$. To generalize, it is necessary to write $\psi(\underline{r})$ under the form

$$\psi(\underline{r}) = \sum_{m=1}^{g} \alpha_m \psi_m(\underline{r}) \tag{2.29}$$

with

$$\psi_m(\underline{r}) = \sum_i \exp(i\underline{k}_{0m} \cdot \underline{R}_i) f_m(\underline{R}_i) W(\underline{r} - \underline{R}_i) \quad . \tag{2.30}$$

This form has the correct limiting behavior, and the functions $f_m(\underline{R}_i)$ all tend towards constant values. We can thus develop a theory similar to the case of one minimum, leading to

$$\sum_m \alpha_m \exp(i\underline{k}_{0m} \cdot \underline{R}_i) \sum_{\underline{R}_h} [H_0(\underline{R}_0, \underline{R}_h) + U(\underline{R}_i, \underline{R}_i + \underline{R}_h)]$$

$$\exp(i\underline{k}_{0m} \cdot \underline{R}_h)f_m(\underline{R}_i + \underline{R}_h) = E \sum_m \alpha_m \exp(i\underline{k}_{0m} \cdot \underline{R}_i)f_m(\underline{R}_i) \qquad (2.31)$$

which is the generalization of (2.15).

We can again use the fact that the $f_m(\underline{R}_i + \underline{R}_h)$ are slowly varying functions. We consider them as being continuous functions and expand them to second order in $\underline{R}_h$. Using one fixed system of coordinates $O\vec{x}$, $O\vec{y}$, $O\vec{z}$, this gives

$$\sum_m \alpha_m \exp(i\underline{k}_{0m} \cdot \underline{x})\left[-\frac{\hbar^2}{2}\sum_{\alpha,\beta}\left(M_m^{-1}\right)_{\alpha\beta}\frac{\partial^2}{\partial x_\alpha \partial x_\beta} + U(\underline{x},\underline{x}) - (E - E_c)\right]f_m(\underline{x}) = 0 \quad ,$$

$$(2.32)$$

whose solutions are to be taken at $\underline{x} = \underline{R}_i$. Here $\left(M_m^{-1}\right)$ is the effective mass tensor at the mth minimum, expressed in the fixed system of coordinates (thus, it is not diagonal in the general case), and E_c is the minimum of the conduction band. Let us call h_m the effective mass hamiltonian for the mth minimum. Multiplying (2.32) on the left by $\exp(-i\underline{k}_{0\ell} \cdot \underline{x})\, f_\ell^*(\underline{x})$ and integrating over $\underline{x}$ gives

$$\sum_m \alpha_m \int \exp[i(\underline{k}_{0m} - \underline{k}_{0\ell})\underline{x}]f_\ell^*(\underline{x})[h_m - (E - E_c)]f_m(\underline{x})d^3\underline{x} = 0 \quad . \qquad (2.33)$$

This is a set of g homogeneous linear equations with g number of unknowns α_m (note that their coefficients contain the $f_m(\underline{x})$ which are also unknown). However, the nondiagonal terms are integrals of the product of $\exp[i(\underline{k}_{0m} - \underline{k}_{0\ell})\underline{x}]$, which is usually a rapidly oscillating function of $\underline{x}$ by a slowly varying function. They are thus very small and this system can be analyzed by conventional first-order perturbation theory. To zeroth order, we only retain the diagonal terms. This leads to

$$[h_m - (E - E_c)]f_m(\underline{x}) = 0 \qquad m = 1,\ldots,g \quad , \qquad (2.34)$$

which represents the effective mass equation for the mth minimum treated independently. Since these minima are equivalent, we end up with a g-fold degenerate level corresponding to the independent wave functions $\psi_m(\underline{x})$ defined in (2.30). Once each $f_m(\underline{x})$ is calculated, it can be injected in (2.33), leading to a completely determined set of equations.

Let us investigate the solution of (2.34) for each minimum considered as being independent of the others. This is the usual effective mass equation, but with the inverse effective mass tensor $M^{-1}_{m_{\alpha\beta}}$ relative to that particular minimum. In the materials of interest (e.g., Si, Ge) symmetry allows complete determination of the set of axes which diagonalize this tensor. For instance, if k_{0m} is along the (100) direction, M_m is diagonal for one axis along that direction, the two other axes being perpendicular to it. Furthermore, the system has cylindrical symmetry. This means that in this set of axes, M^{-1} becomes

$$M^{-1} = \begin{bmatrix} 1/m^*_{\ell} & & 0 \\ & 1/m^*_t & \\ 0 & & 1/m^*_t \end{bmatrix} \quad ; \tag{2.35}$$

m^*_{ℓ} and m^*_{ℓ} are respectively the longitudinal and transverse effective masses. Equation (2.34) then becomes

$$\left\{ -\frac{\hbar}{2} \left[\frac{1}{m^*_{\ell}} \frac{d^2}{dx^2} + \frac{1}{m^*_t} \left(\frac{d^2}{dy^2} + \frac{d^2}{dz^2} \right) \right] + U(\underline{x},\underline{x}) \right\} f_m(\underline{x}) = (E - E_c) f_m(\underline{x}) \quad . \tag{2.36}$$

This equation can be solved by the method of variations with a trial function having cylindrical symmetry. Here we only want an order of magnitude estimate, so that we make a spherical average, which has the simple form found in Sect.2.2.1, with

$$\frac{1}{m^*} = \frac{1}{3} \left(\frac{1}{m^*_{\ell}} + \frac{2}{m^*_t} \right) \quad . \tag{2.37}$$

The results will be compared to experiment in Sect.2.4.

Once the zeroth-order approximation to $f_m(\underline{x})$ given by the solution of (2.34) is known, we must come back to (2.33). The solution proceeds exactly as in first-order perturbation theory on a g-fold degenerate state. The non-diagonal terms will partly lift this degeneracy. This fact is observed experimentally.

2.2.3 Validity of the Approximations

In our discussion so far, different approximations have been made which are not independent of one another. One of them consists in the neglect of the terms $U(\underline{R}_i, \underline{R}_j)$ for $R_i \neq R_j$. This is valid when $U(\underline{r})$ is practically constant over one unit cell, i.e., when its variation ΔU over one lattice constant a_0 is small compared to the value of the potential $U(\underline{r})$ itself. For a Coulomb potential, this is verified when $a_0/r \ll 1$ i.e., at large distances from the impurity. The total error due to this approximation depends on the relative contributions to the energy from regions close to the impurity site. This error will become smaller for more extended wave functions. For the ground state wave function, which behaves as $\exp(-r/a)$, this means that the Bohr radius a must be much greater than the interatomic distance i.e., $a \gg a_0$. This condition will be better satified for small effective masses and large dielectric constants. It will also be more easily obeyed for excited states whose wave functions are more extended in space.

The other approximation consists in retaining only second-order terms in the expansion of $f(\underline{x} + \underline{R}_h)$. This is qualitatively valid only in the limit of a very smooth function $f(\underline{x})$, i.e., a function which does not vary much over a unit cell, which again corresponds to the mathematical condition $a \gg a_0$. Finally, the fact of retaining only one band in the calculation leads to errors that can be estimated by perturbation theory, but are weak since the matrix elements of $U(\underline{r})$ between Wannier functions belonging to different bands should be small in view of their orthogonality.

We shall see in Sect.2.4 that in usual cases, the most important errors introduced by the effective mass theory occur for the ground state, as expected. An important part of this error comes from the deviation of the true potential $U(\underline{r})$ from its Coulomb expression. This deviation occurs near the impurity and is known as "central cell correction". Again it is more important for the ground state. There is also a correction due to the fact that screening is not so efficient near the impurity site. This can be accounted for by the use of a wave vector or of a position-dependent dielectric constant. A final and important cause of errors in the ground-state binding energy comes from the splitting induced by the interaction between different minima discussed in Sect.2.2.2, i.e., intervalley mixing.

2.2.4 Generalization to the Case of a Degenerate Extremum

The top of the valence band at $\underline{k} = 0$ in zinc blende semiconductors has threefold degeneracy when the spin-orbit coupling is neglected. In such a case it is not possible to expand the energy $E_n(\underline{k})$ in powers of $\underline{k}$ (n is the band index). The previous derivation then fails since it involves second-order derivatives $\left(\partial^2 E_n/\partial k_\alpha \partial k_\beta\right)_{k=0}$. We then proceed in this section to a new derivation of the effective mass approximation which directly applies to this case.

Let us label $\underline{k}_0$ the wave vector of the degenerate extremum. As in Sect. 2.2.1 we would like to express the localized-impurity-state wave function $\psi(\underline{r})$ in the basis of the unperturbed-crystal wave functions. In this section we have used Wannier functions as basis states. We could also have used Bloch functions $\psi_{n\underline{k}}(\underline{r})$, but with the same difficulties in applying the method to the degenerate case. To avoid these difficulties it is necessary to use a new set of basis functions around the degenerate extremum at $\underline{k}_0$. These functions, which have been introduced by LUTTINGER and KOHN [2.15] are defined as

$$\Phi_{n\underline{k}}(\underline{r}) = \exp[i(\underline{k} - \underline{k}_0)\underline{r}]\psi_{n,\underline{k}_0}(\underline{r}) \quad . \tag{2.38}$$

It can be shown [Ref.2.5, p.217] that they form an orthonormal basis set, exactly as the Bloch functions $\psi_{n\underline{k}}(\underline{r})$, but their use will remove the mathematical difficulties associated with the degeneracy. However, instead of working directly with the $\Phi_{n\underline{k}}$, we shall again take their Fourier transforms as we have done previously for the $\psi_{n\underline{k}}$. We thus define modified Wannier functions $v_n(\underline{r} - \underline{R}_i)$ as

$$v_n(\underline{r} - \underline{R}_i) = \frac{1}{\sqrt{N}} \sum_{\underline{k}} \exp(-i\underline{k} \cdot \underline{R}_i)\Phi_{n\underline{k}}(\underline{r}) \quad . \tag{2.39}$$

The set of $v_n(\underline{r} - \underline{R}_i)$ defined in this way is also orthonormal since the quantities $(1/\sqrt{N}) \exp(-\underline{k}\underline{R}_i)$ are the coefficients of a unitary transformation. These modified Wannier functions have also a localized character since they are Fourier transforms of delocalized functions. We can thus expand $\psi(\underline{r})$, as we have done before, in terms of this new set of basis functions

$$\psi(\underline{r}) = \sum_{m,j} \exp(i\underline{k}_0 \cdot \underline{R}_j)f_m(\underline{R}_j)v_m(\underline{r} - \underline{R}_j) \quad , \tag{2.40}$$

where the index m is summed over the different bands. We can write the analog of (2.14) by projecting Schrödinger's equation onto one basis function $v_n(\underline{r} - \underline{R}_i)$. This leads to

$$\sum_{m,\underline{R}_j} <v_n(\underline{r} - \underline{R}_i)|H_0 + U(\underline{r})|v_m(\underline{r} - \underline{R}_j)> \exp(i\underline{k}_0 \cdot \underline{R}_j)f_m(\underline{R}_j) = Ef_n(\underline{R}_i) \quad .$$

$$\tag{2.41}$$

Contrary to the true Wannier functions, the Hamiltonian H_0 now has matrix elements between $v_n(\underline{r} - \underline{R}_i)$ belonging to different bands ($m \neq n$) since the v_n

are combinations of Φ_{nk} which are not eigenstates of H_0 except at $\underline{k} = \underline{k}_0$. In the limit of slowly varying $U(\underline{r})$ we can again write

$$<v_n(\underline{r} - \underline{R}_i)|U(\underline{r})|v_m(\underline{r} - \underline{R}_j)> = U(\underline{R}_i,\underline{R}_i)\delta_{mn}\delta_{ij} \quad . \tag{2.42}$$

Using the same arguments as before, we transform (2.42) to obtain

$$\sum_{\underline{R}_h} \sum_m <v_n(\underline{r} - \underline{R}_0)|H_0|v_m(\underline{r} - \underline{R}_h)> \exp(i\underline{k}_0 \cdot \underline{R}_h)f_m(\underline{x} + \underline{R}_h)$$

$$+ U(\underline{x})f_n(\underline{x}) = Ef_n(\underline{x}) \quad , \tag{2.43}$$

where $\underline{x}$ is now treated as a continuous variable. The main problem is reducing the first term in (2.43) and showing that it leads to a generalization of the effective mass approximation. For this we use the equality

$$<\Phi_{nk}|H_0|\Phi_{mk}> = \sum_{\underline{R}_h} \exp(i\underline{k} \cdot \underline{R}_h) <v_n(\underline{r} - \underline{R}_0)|H_0|v_m(\underline{r} - \underline{R}_h)> \quad , \tag{2.44}$$

which simply results from the definition of the $v_n(\underline{r} - \underline{R}_i)$ and from the translational invariance of H_0. Before detailing the matrix element occuring in (2.41), we complete the derivation of the effective mass equation. We again assume in (2.43) that $f_m(\underline{x} + \underline{R}_h)$ is a slowly varying function so that we can expand it to second order in $\underline{R}_h$. The first member of (2.43) can be rewritten as

$$\sum_{\underline{R}_h} \sum_m <v_n(\underline{r} - \underline{R}_0)|H_0|v_m(\underline{r} - \underline{R}_h)> \exp(i\underline{k}_0 \cdot \underline{R}_h)$$

$$\left[1 + \sum_\alpha x_h^\alpha \frac{\partial}{\partial x_\alpha} f_m(\underline{x}) + \frac{1}{2} \sum_{\alpha\beta} x_h^\alpha x_h^\beta \frac{\partial^2}{\partial x_\alpha \partial x_\beta} f(\underline{x}) \dots \right] \quad , \tag{2.45}$$

which is formally identical to

$$\sum_m \left(1 + \frac{1}{i} \sum_\alpha \frac{\partial}{\partial x_\alpha} \cdot \frac{\partial}{\partial k_\alpha} - \frac{1}{2} \sum_{\alpha\beta} \frac{\partial^2}{\partial k_\alpha \partial k_\beta} \frac{\partial^2}{\partial x_\alpha \partial x_\beta}\right) <\Phi_{nk}|H_0|\Phi_{mk}>f_m(\underline{x}) \quad , \tag{2.46}$$

where the derivatives with respect to $\underline{k}$ have to be taken at $\underline{k}_0$. From the definition of the Φ_{mk} given by (2.38), it is already evident that there is no mathematical problem in taking these derivatives.

Let us now investigate the detailed form of each matrix element $<\Phi_{nk}|H_0|\Phi_{mk}>$ (in the absence of spin-orbit coupling). For this let us notice that H_0 is the sum of a kinetic energy term $p^2/2m$ and of a potential energy V_0. The action of H_0 on Φ_{mk} can be written

$$\left(\frac{p^2}{2m} + V_0\right) \exp[i(\underline{k} - \underline{k}_0)\underline{r}]\psi_{m\underline{k}_0}(\underline{r}) = \exp[i(\underline{k} - \underline{k}_0)\underline{r}]$$

$$\left(\frac{[\underline{p}+\underline{\hbar}(\underline{k}-\underline{k}_0)]^2}{2m} + V_0\right)\psi_{m\underline{k}_0}(\underline{r}) \quad , \tag{2.47}$$

so that we obtain for the matrix element

$$\langle\Phi_{n\underline{k}}|H_0|\Phi_{m\underline{k}}\rangle = \langle\psi_{n\underline{k}_0}|H_0 + \frac{\hbar}{m}(\underline{k} - \underline{k}_0)\cdot\underline{p} + \frac{\hbar^2(k-k_0)^2}{2m}|\psi_{m\underline{k}_0}\rangle \quad , \tag{2.48}$$

that is,

$$\langle\Phi_{n\underline{k}}|H_0|\Phi_{m\underline{k}}\rangle = \left[E_m(\underline{k}_0) + \frac{\hbar^2}{2m}(\underline{k} - \underline{k}_0)^2\right]\delta_{m,n} + \frac{\hbar}{m}(\underline{k} - \underline{k}_0)\langle\psi_{n\underline{k}_0}|\underline{p}|\psi_{m\underline{k}_0}\rangle \quad . \tag{2.49}$$

The derivatives with respect to $\underline{k}$ that occur in (2.46) can now be evaluated, leading to

$$\sum_m \left[E_m(\underline{k}_0)\delta_{mn} - \frac{i\hbar}{m}\langle\psi_{n\underline{k}_0}|\underline{p}|\psi_{m\underline{k}_0}\rangle\cdot\underline{\nabla} - \delta_{mn}\cdot\frac{\hbar^2}{2m}\Delta\right]f_m(\underline{x}) \quad , \tag{2.50}$$

where $\underline{\nabla}$ and Δ are the gradient and Laplacian operators. Introducing (2.50) into (2.41), the effective mass equation then becomes

$$\left[-\frac{\hbar^2}{2m}\Delta + U(\underline{x})\right]f_n(\underline{x}) - \frac{i\hbar}{m}\sum_m\langle\psi_{n\underline{k}_0}|\underline{p}|\psi_{m\underline{k}_0}\rangle\underline{\nabla}f_m(\underline{x}) = [E - E_n(\underline{k}_0)]f_n(\underline{x}) \quad . \tag{2.51}$$

Clearly, this is a set of coupled differential equations, even near a non-degenerate extremum, since now all bands contribute. Fortunately, this set of equations can be solved approximately if we search for solutions having an energy E close to one given extremum $E_v(\underline{k}_0)$ whose possible Bloch functions are $\psi_{v_i}(\underline{k}_0)$ (i running over the degenerate set). In this case we separate the equations corresponding to $n = v_i$ from the other equations. We thus write

$$\left[-\frac{\hbar^2}{2m}\Delta + U(\underline{x})\right]f_{v_i}(\underline{x}) - \frac{i\hbar}{m}\sum_{m\neq v_i}\langle\psi_{v_i\underline{k}0}|\underline{p}|\psi_{m\underline{k}0}\rangle\underline{\nabla}f_m(\underline{x})$$

$$= [E - E_v(\underline{k}_0)]f_{v_i}(\underline{x}) \quad . \tag{2.52}$$

Due to symmetry, there are no matrix elements of $\underline{p}$ within the degenerate set in this equation. We also write the equation for values of n different from v_i

$$\left[E - E_n(\underline{k}_0) + \frac{\hbar^2}{2m}\Delta - U(\underline{x})\right]f_n(\underline{x}) = \frac{i\hbar}{m}\sum_{m\neq n}<\psi_{n\underline{k}_0}|\underline{p}|\psi_{m\underline{k}_0}> \underline{\nabla}f_m(\underline{x}) \quad . \quad (2.53)$$

Because $n \neq v_i$, the quantity $E - E_n(\underline{k}_0)$ is very close to $E_v(\underline{k}_0) - E_n(\underline{k}_0)$ since the bound-state level which we are looking at is close to $E_v(\underline{k}_0)$. The contribution of $\hbar^2\Delta/2m - U(\underline{x})$ acting on the smooth function $f_n(\underline{x})$ will be very small so we will neglect it. We can then write approximatively

$$f_n(\underline{x}) = \frac{i\hbar}{m}\sum_{m\neq n}\frac{<\psi_{n\underline{k}_0}|\underline{p}|\psi_{m\underline{k}_0}>}{E_v(\underline{k}_0)-E_n(\underline{k}_0)}\underline{\nabla}f_m(\underline{x}) \quad . \quad (2.54)$$

This result is equivalent to the result given by first-order perturbation theory on the wave functions. Injecting this expression of $f_n(\underline{x})$ into (2.52) and retaining terms to the second order in the ratio $<\psi_{n\underline{k}_0}|\underline{p}|\psi_{m\underline{k}_0}>$ over $E_v(\underline{k}_0) - E_n(\underline{k}_0)$, we obtain

$$\left[-\frac{\hbar^2}{2m}\Delta + U(\underline{x})\right]f_{v_i}(\underline{x})$$

$$+ \frac{\hbar^2}{m^2}\sum_{\substack{v_j \\ v_j}}\sum_{\substack{m\neq v_i \\ v_j}}\frac{\left(<\psi_{v_i\underline{k}_0}|\underline{p}|\psi_{m\underline{k}_0}>\cdot\underline{\nabla}\right)\left(<\psi_{m\underline{k}_0}|\underline{p}|\psi_{v_j\underline{k}_0}>\cdot\underline{\nabla}\right)}{E_v(\underline{k}_0)-E_m(\underline{k}_0)}f_{v_j}(\underline{x})$$

$$= [E - E_v(\underline{k}_0)]f_{v_i}(\underline{x}) \quad (2.55)$$

Equation (2.55) is the useful form of the effective mass approximation in the degenerate case. Furthermore, it gives a method for the calculation of the effective mass. This can be easily seen in the nondegenerate case. In such a case there is only one function $f_v(\underline{x})$ corresponding to $E_v(\underline{k}_0)$ and (2.55) reduces to the usual one-band expression at the condition of writing the inverse effective mass tensor as

$$M^{-1}_{\alpha\beta} = \frac{1}{m}\delta_{\alpha\beta} + \frac{2}{m^2}\sum_{m\neq v}\frac{<\psi_{v\underline{k}_0}|p_\alpha|\psi_{m\underline{k}_0}><\psi_{m\underline{k}_0}|p_\beta|\psi_{v\underline{k}_0}>}{E_m(\underline{k}_0)-E_v(\underline{k}_0)} \quad . \quad (2.56)$$

This expression is equivalent to second-order $\underline{k} \cdot \underline{p}$ perturbation theory [2.16]. The generalization provided by (2.55) leads to an effective mass matrix within the degenerate manifold whose general element $\left(M^{-1}\right)^{ij}_{\alpha\beta}$ is equal to

$$\left(M^{-1}\right)_{\alpha\beta}^{ij} = \frac{1}{m}\,\delta_{\alpha\beta}\delta_{ij} + \frac{2}{m^2}\sum_{m\neq v_i,v_j}\frac{<\psi_{v_i\underline{k}0}|p_\alpha|\psi_{m\underline{k}0}><\psi_{m\underline{k}0}|p_\beta|\psi_{v_j\underline{k}0}>}{E_m(\underline{k}_0)-E_v(\underline{k}_0)} \tag{2.57}$$

Finally, the effective mass equation (2.55) can be written

$$-\frac{\hbar^2}{2}\sum_j\sum_{\alpha,\beta}\left(M^{-1}\right)_{\alpha\beta}^{ij}\frac{\partial^2}{\partial x_\alpha\,\partial x_\beta}f_{v_j}(\underline{x}) + U(\underline{x})f_{v_i}(\underline{x}) = [E - E_v(\underline{k}_0)]f_{v_i}(\underline{x}) \quad. \tag{2.58}$$

This system of coupled differential equations is much more difficult to solve than in the nondegenerate case. Usually this is done by means of a variational procedure.

In practice for silicon and germanium the problem is complicated by spin-orbit coupling which splits the degenerate level and changes the definition of the effective mass. However, we can use exactly the same theory as before, simply changing p into a more complicated term [Ref.2.5, p.255]

$$\underline{p} + \frac{\hbar}{4m^2c^2}\,\underline{\sigma}\wedge\underline{\nabla}V \quad, \tag{2.59}$$

where $\underline{\sigma}$ is the spin and V the potential; $\wedge$ stands for the vector product. In particular the effective mass matrix keeps exactly the same form as before, once the components of $\underline{p}$ are replaced by the components of the operator (2.59).

The effective mass matrix defined in (2.57) allows the determination of the energy band shapes in the vicinity of the degenerate extremum $E_v(\underline{k}_0)$ since it corresponds to the expansion of terms like $<\phi_{n\underline{k}}|H_0|\phi_{m\underline{k}}>$, as shown previously. To obtain these energy bands we simply have to diagonalize the matrix of general element

$$\sum_{\alpha\beta}\left(M^{-1}\right)_{\alpha\beta}^{ij}\left(\underline{k} - \underline{k}_0\right)_\alpha\left(\underline{k} - \underline{k}_0\right)_\beta \quad. \tag{2.60}$$

Each individual matrix element corresponds to a second-order expansion in powers of $\underline{k}$, but this does not mean that the eigenvalues can be expanded in terms of this quantity. For instance, the diagonalization, when spin-orbit coupling is included, yields for the three highest valence bands of silicon and germanium [Ref.2.5, p.161] to

$$E_{12} = A\underline{k}^2 \pm \left[B^2\underline{k}^4 + C^2\left(k_x^2k_z^2 + k_y^2k_z^2 + k_x^2k_y^2\right)\right]^{\frac{1}{2}} \quad, \tag{2.61}$$

$$E_3 = -\Delta + Ak^2 \quad , \tag{2.62}$$

where Δ is the spin-orbit splitting and A,B,C are parameters related to the $\left(M^{-1}\right)_{\alpha\beta}^{ij}$ whose number can be reduced by symmetry arguments. Clearly, E_{12} is not expandable in powers of $\underline{k}$, meaning that a simple direct application of effective mass theory is not valid.

2.3 Pairing Effects

When a semiconductor contains at the same time donor and acceptor defects in concentrations large enough so that their mutual interaction is not negligible, the binding energies of the corresponding electron, hole, or electron-hole pair is modified. The change in binding energy is a function of the distance between impurities. Since there is a distribution of distances, there is a distribution of binding energies which can provide informations on the statistical distribution of the impurity pairs.

2.3.1 An Electron Bound to a Donor-Acceptor Pair

The first case, illustrated in Fig.2.3, corresponds to an electron bound at a donor-acceptor pair. The electron bound to the donor A is perturbed by the presence of the ionized acceptor B. As discussed above, the effective mass theory allows the reduction of this problem to the one of an hydrogen atom (electron + nucleus) in the field of a point negative charge localized at nucleus B situated at a distance R from A (we ignore complications arising from anisotropy and equivalent minima). Let us consider this problem for $m_n^* = m$, $\varepsilon = 1$ (the results can be scaled at the end, replacing m by m_n^* and e^2 by e^2/ε).

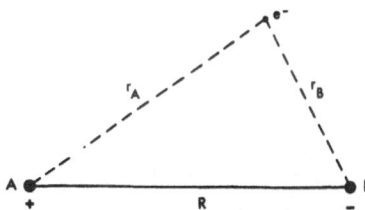

Fig.2.3. Electron bound to a donor (A) - acceptor (B) pair

54

There are, in principle, two limits in which this problem can be simplified. The first one corresponds to $R > a_0$ the Bohr radius of the hydrogen
atom. Then the effect of B can be treated by first-order perturbation theory.
The second limit corresponds to small distances R and could be treated in
the point-dipole approximation. This limit is not very simple to analyze.
There are not always bound states and the binding energies are very small
[2.17]. For this reason we only examine the results starting from the first
limiting situation.

Exact calculations have been performed using spheroidal coordinates in
which the Schrödinger equation separates. The results for the ground state
are given in Fig.2.4. In this figure are also given the values obtained by
perturbation theory, the change in the binding energy δE being given by

$$\delta E = <\psi_{1s}|-e^2/r_B|\psi_{1s}> \quad . \tag{2.63}$$

where r_B is the distance between the electron and the acceptor and ψ_{1s} is the
1s ground-state wave function for the isolated hydrogen atom. A straightforward calculation [2.18] leads to the result

$$\delta E = \frac{-e^2}{R}\left[1 - \left(1 + \frac{R}{a_0}\right)\exp\left(-\frac{2R}{a_0}\right)\right] \quad . \tag{2.64}$$

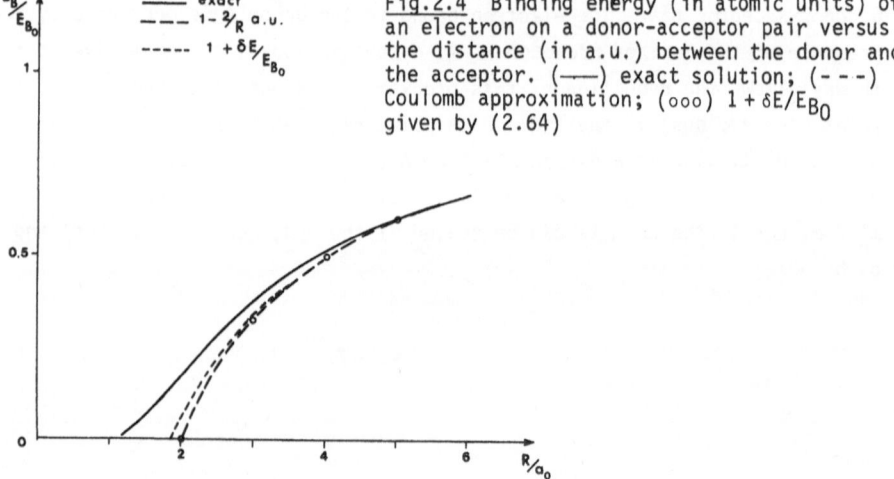

Fig.2.4 Binding energy (in atomic units) of
an electron on a donor-acceptor pair versus
the distance (in a.u.) between the donor and
the acceptor. (——) exact solution; (- - -)
Coulomb approximation; (ooo) $1 + \delta E/E_{B_0}$
given by (2.64)

The meaning of the first term is obvious: this is the potential created by B
on the electron at site A. The second term gives the deviations to this potential due to the spreading out of the electron distribution. In Fig.2.4 we

have plotted the corresponding two curves, i.e., $1 - e^2/RE_{B_0} = 1 - 2 a_0/R$
or $1 - 2/R$ in atomic units (where E_{B_0} is the isolated hydrogen atom binding
energy) and $1 + \delta E/E_{B_0}$. Comparison with the exact curve shows that the simplest
correction $-e^2/R$ for the binding energy accounts for most of the effect and
that the exponential part of (2.64) only gives a minor correction. Obviously,
any serious improvement requires allowance for a polarization of the electron
wave function along the AB axis. However, this presents no interest for the
discussion which follows.

The results presented in Fig.2.4 can be applied to donor-acceptor pairs
in any case where effective-mass theory is applicable. The only changes which
are required consist in the scaling of a_0 from its hydrogen value to its actual value $\epsilon m/m^*$ (in atomic units), and of E_{B_0} from 1 Rydberg to $m^*/(m\epsilon^2)$
Rydberg units. A very important point which is apparent in Fig.2.4 is that
the binding energy is lowered to about one quarter of its original value for
$R/a_0 \simeq 2$. Taking the example of silicon ($m^* \simeq 0.25$ and $\epsilon \simeq 10$, giving
$a_0 \simeq 20$ Å), this means that for $R \simeq 40$ Å the binding energy of the ground
state will be of the same order of magnitude as the energy of the p states
of the isolated donor. This distance is far greater than the nearest neighbor's distance in the silicon lattice (2.35 Å), meaning that near neighbors'
pairs will not be observable in that case. Only in systems where a_0 is much
smaller will this conclusion no longer hold, but in such systems the effective mass approximation is not valid and can only be used to get rough
estimations.

2.3.2 The Neutral Donor-Acceptor Pair

We now discuss the transition energies corresponding to photo-excitation and
radiative deexcitation at neutral donor-acceptor pairs. This transition energy
ΔE is calculated as follows. a) Create a free electron-hole pair; this requires an energy equal to the energy gap E_g. b) Calculate the gain in energy,
or binding energy E_B, obtained by trapping the electron and the hole at the
donor-acceptor pair. The transition energy ΔE then comes out to be

$$\Delta E = E_g - E_B \quad . \tag{2.65}$$

Again using the effective mass theory, the problem reduces to the one schematized in Fig.2.5. A detailed theory is again difficult but we can, as previously, work out the situation in the large distance limit. We start from a
neutral donor ($A + e^-$) and a neutral acceptor ($B + h^+$) that interact weakly so
that we can use a first-order perturbation theory. The unperturbed wave func-

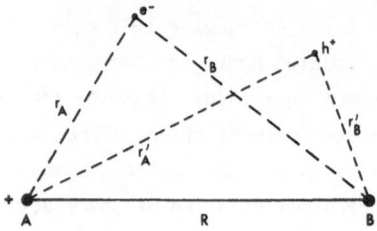

Fig.2.5. Electron-hole pair bound to a donor (A) - acceptor (B) pair (the neutral A-B pair)

tion ψ, corresponding to the noninteracting species is simply given by the product of the individual wave functions $\phi_A(\underline{r})$ and $\phi_B(\underline{r}')$:

$$\psi = \phi_A \cdot \phi_B \quad , \tag{2.66}$$

$\underline{r}$ and $\underline{r}'$ being the vector positions of the electron and hole. The unperturbed binding energy is the sum of the donor E_D and acceptor E_A binding energies given by the usual hydrogenic formula. The correction introduced by first-order perturbation theory is the average value $<V>$ of the perturbative potential

$$<V> = \left\langle \phi_A\phi_B \left| \frac{e^2}{r_B} + \frac{e^2}{r_A'} - \frac{e^2}{r_{eh}} \right| \phi_A\phi_B \right\rangle \quad , \tag{2.67}$$

or

$$<V> = \int d^3\underline{r}\,d^3\underline{r}' \,|\phi_A(\underline{r})|^2 |\phi_B(\underline{r}')|^2 \left(\frac{e^2}{r_B} + \frac{e^2}{r_A'} - \frac{e^2}{r_{eh}} \right) \quad . \tag{2.68}$$

At distances R large compared to the Bohr radii of both donor and acceptor, each term in (2.68) will give a contribution of the order of e^2/R, so that $<V>$ will also tend towards this value. To have an idea of the order of magnitude of the correction, we write $<V>$ as $(e^2/R) + J$ and evaluate J assuming equal effective masses for the donor and the acceptor (in which case the integrals are exactly the same as for the hydrogen molecule). The evaluation of J [2.18] leads to

$$J = \frac{e^2}{R} \left[-1 - \frac{5}{8}\frac{R}{a_0} + \frac{3}{4}\left(\frac{R}{a_0}\right)^2 + \frac{1}{6}\left(\frac{R}{a_0}\right)^3 \right] \exp\left(-\frac{2R}{a_0}\right) \quad . \tag{2.69}$$

With the above notations, the binding energy E_B becomes equal to

$$E_B = 2E_{B0} - \frac{e^2}{R} - J \quad . \tag{2.70}$$

The different contributions to E_B/E_{B0} are plotted in Fig.2.6. Again it is obvious that the correction introduced by J is of very small importance so that over most of the interesting range, the e^2/R term gives the major part of the interaction energy. However it is clear that the behavior for small R is incorrect since generally in that limit, E_B must tend to the free exciton binding energy. An abrupt change in the curve should occur about $R/a_0 \simeq 1.4$ [2.19]. Thus for $R > 2 \, a_0$, we can consider that (2.70) is valid. It can be scaled for m^* and ε and leads to the well known formula

$$E_B = E_D + E_A - \frac{e^2}{\varepsilon R} \quad . \tag{2.71}$$

In ionic systems complications arise due to the polarization of the lattice, but we shall not consider such effects here.

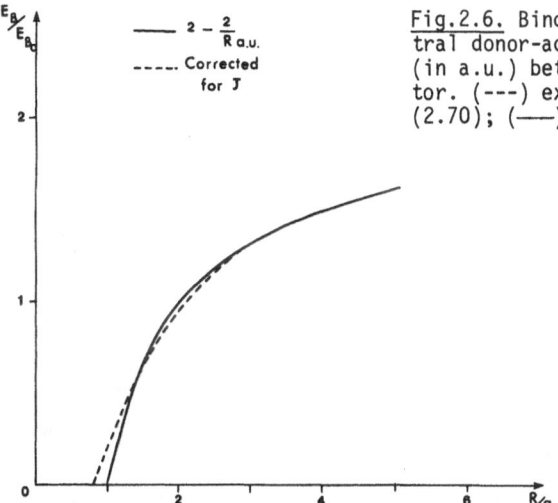

Fig.2.6. Binding energy (in a.u.) on a neutral donor-acceptor pair versus the distance (in a.u.) between the donor and the acceptor. (---) exact variation according to (2.70); (—) neglecting the J term

2.3.3 Density of States

The distribution of donor-acceptor pairs in a crystal results from an equilibrium at the temperature T_0 corresponding to crystal preparation, which is much higher than the temperature T at which the properties of the material are usually studied. Consider, for instance, the case where the defect, which is mobile at T_0, is the donor. We define the pair distribution function g(R) as the probability per unit length of finding the nearest donor at distances lying between R and $R + dR$ of a given acceptor. Statistical theory (see Sect. 6.5) shows that in many cases g(R) is a two peaked function which can in practice be written as the sum of two contributions $g_1(R)$ and $g_2(R)$, as shown in Fig.2.7.(In this figure R_0 is the distance of closest approach between a

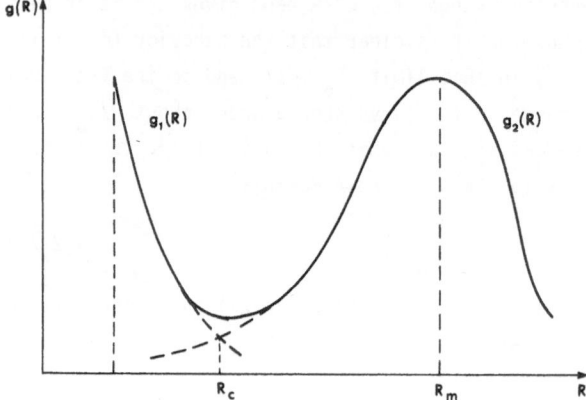

g(R)

g₁(R)

g₂(R)

R_c R_m R

<u>Fig.2.7.</u> Pair distribution function, i.e., number of pairs separated by a distance R, versus R

donor and an acceptor, R_c and R_m define the positions of the minimum and of the maximum of the distribution.) The figure assumes that R is a continuous variable. In some cases the resolution is such that it is necessary to take account of the discreteness of R. However, the general features of g(R) remain the same.

An important parameter is the distance R_c, which separates the two distributions. In silicon $(\varepsilon \simeq 10)$ for impurities migrating at 1000 K, (6.93) gives for R_c a value of 8 Å, which is smaller than the Bohr radius of the donor. This has implications on the experimental results since we have seen that the pair spectra depend on the pair separation R and that the binding energy $E_B(R)$ is in all cases vanishing small for $R \lesssim a_0$. Sizable binding energies will be obtained only for R much greater than R_c and their experimental observation will then reflect the distribution $g_2(R)$.

Let us investigate the shape of the density of states. Formally, the transition energy is some function of the pair binding energy $E_B(R)$, which depends upon the distance. The density of states $n(E_B)$ can be calculated as follows. The number of defects with energy in the range between E_B, $E_B + dE_B$ is $n(E_B)dE_B$. Since E_B is a function of R, $n(E_B)dE_B$ is equal to the number of defects in the range $R(E_B)$ to $R(E_B) + dR$, where $R(E_B)$ is the pair distance corresponding to a given binding energy. We can therefore write the normalized density of states

$$n(E_B) = g_2(R) \frac{dR}{dE_B} \quad , \qquad (2.72)$$

where $g_2(R)$ is the normalized pair distribution function. We will show in Sect.6.6 that $g_2(R)$, which corresponds to a random distribution of donors (in concentration on N_D) around the acceptors, is given by

$$g_2(R) \simeq 4\pi R^2 N_D \exp\left(- \frac{4\pi}{3} R^3 N_D\right) \quad . \tag{2.73}$$

In that case taking into account the fact that in this range E_B can be approximated by

$$E_B = E_{B0} - \frac{e^2}{\varepsilon R} \quad , \tag{2.74}$$

we find

$$n(E_B) = 4\pi R^2 N_D \frac{\varepsilon R^2}{e^2} \exp\left(- \frac{4\pi}{3} R^3 N_D\right) \quad , \tag{2.75}$$

or equivalently

$$n(E_B) = 4\pi \left(\frac{e^2}{\varepsilon}\right)^3 N_D \frac{1}{(E_{B0}-E_B)^4} \exp\left[- \frac{4\pi}{3} \left(\frac{e^2}{\varepsilon}\right)^3 \frac{N_D}{(E_{B0}-E_B)^3}\right] \quad . \tag{2.76}$$

The logarithmic derivative of this quantity is

$$\frac{d}{dE_B} \log n(E_B) = + \frac{4}{E_{B0}-E_B} - 4\pi \left(\frac{e^2}{\varepsilon}\right)^3 \frac{N_D}{(E_{B0}-E_B)^4} \quad . \tag{2.77}$$

This expression tends to zero when E_B tends to E_{B0}. It reaches a maximum for

$$E_{B0} - E_B = \frac{e^2}{\varepsilon} (\pi N_D)^{1/3} \quad . \tag{2.78}$$

The shape of $n(E_B)$ is given in Fig.2.8.

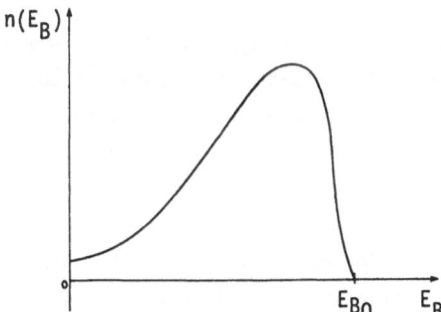

Fig.2.8.Density of states associated with a pair distribution

2.4 Experimental Observation of Shallow Levels

2.4.1 Experimental Techniques

The way to investigate the position of a localized level in a forbidden gap
is to observe an electronic transition between this level and a band. The
transition can be induced by an electromagnetic wave (optical transition) or
thermally. The observation of emission or absorption [Ref.1.1, Chap.10] of
excitation radiation and the variation of the free carrier concentration versus time at a given temperature [Ref.1.1, Chap.11] or versus the wave length
of the excitation give a measure of the position of the localized level. Additional information on the levels can be obtained from the hyperfine interaction detected by electron paramagnetic resonance [Ref.1.1, Chap.9]. The
interaction of the spin of the electron with the magnetic moments of the
nuclei present within its orbit allows the mapping of the wave function;
this interaction provides the amplitude of the wave function at the position
of these nuclei (which can be impurities as well as the atoms of the crystal
itself when they possess a nuclear spin).

For shallow levels we have shown that the simplest theory (corresponding
to an isotropic nondegenerate extremum) predicts a ground state at

$$E_{1s} = 13.6 \frac{m^*}{m \epsilon^2} \ [eV] \tag{2.79}$$

from the band, a value which typically varies from a few meV to 300 meV (m^*
ranging from 0.01 m to m and ϵ from 5 to 16). When using optical absorption
measurements, the number of excited states which are observable depend on the
linewidth ΔE, (due to the interaction of electrons with acoustic phonons) as
compared to the energy separation of two adjacent excited states n and n + 1

$$\Delta E \leq E_{1s}\left(\frac{1}{n^2} - \frac{1}{(n+1)^2} \right) \quad . \tag{2.80}$$

At liquid helium temperature, this linewidth is usually ≤ 0.1 meV. This means
that for an impurity whose ground state is at ~10 meV, it is possible to observe transitions to levels with n = 5 to 6 (using an interferometer). The
order of magnitude of the optical energy corresponds to a wavelength for the
electromagnetic radiation in the near infrared range ($\lambda > 10$ µm in the case
of silicon and in the far infrared range ($\lambda > 50$ µm in case of germanium).

Photoconductivity experiments, in which the optically induced transition
is detected as a threshold in the free carrier concentration (measured by
conductivity), are usually not sensitive enough for the observation of excited

states. However, when the optical excitation induces a transition from the ground to an excited state which is then thermally excited in the band at low temperature, this process (schematized in Fig.2.9) leads to sharp lines in the conductivity spectrum [2.20]. For instance, using this technique, it has been possible to observe [2.21] in germanium the isotope shift of the ground state of a shallow defect containing hydrogen or deuterium.

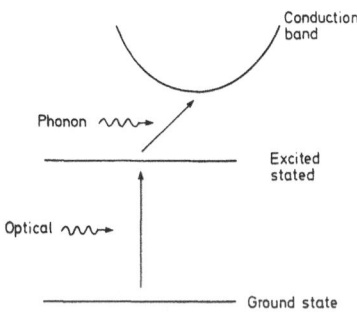

Fig.2.9. Schematic representation of the photothermal ionization process

2.4.2 Results

The experimental values obtained for the ground states for Group IV and two Group III-V semiconductors are compared in Table 2.1 with the calculated ones, using the simple formula given by (2.79). Obviously, this simple formula does not give the right answer for Group IV semiconductors. This is also illustrated by the fact that the experimental value of the binding energy in these materials depends on the nature of the impurity itself (see Table 2.2). A reasonable agreement between the experimental and calculated values is obtained only for GaAs and InP. The reason is that for these materials, the effective mass is isotropic. The simple formula (2.79) implies an isotropic wave function $\psi_{n\ell}$ for the ground state. For orbitally degenerate states, i.e., for states corresponding to an orbital quantum number ℓ larger than 1, the anisotropy of the effective mass induces the splitting of the states having different values of their azimuthal quantum number $m_{\ell}(-\ell < m < +\ell)$. Such spliting is illustrated in Fig.2.10 for the case of 2p($n = 2$, $\ell = 1$) and 3p ($n = 3$, $\ell = 1$) states. When this splitting is large enough (in the case of germanium, for instance) it can result in a $3p_0$ level lying below a $2p \pm 1$ level. Experimentally, (see Table 2.3) the energy levels agree well with the theory for $\ell > 1$, i.e., for excited states. But for the ground state (1s), there is still

62

Table 2.1. Calculated and experimental values of the binding energy (meV) of
s states in various semiconductors. In the case of diamond, the impurity considered is of acceptor type (boron) and because there is a large uncertainty
on the value of m* observed for holes, we have taken arbitrarily an average
value. For the other elements we have considered substitutional donor impurities (taking for m* the effective mass of electrons)

| Material | m_e/m | m_t/m | m^*/m | ε | Calculated | | Experimental | |
					E_{1s}	E_{2s}	E_{1s}	E_{2s}
Diamond	-	-	0,7	5,7	290	73	370	-
Silicon	0,98	0,19	0,26	12	24	6	43,1-53.7	-
Germanium	1,60	0,08	0,12	16	6,6	1,6	10,3-14,2	-
Gallium-arsenide	0,066	0,066	0,066	12,5	5,74	1,43	5,89-6,08	1,44
Indium-phosphide	0,080	0,080	0,080	12,6	6,85	1,71	7,28	-

Table 2.2. Experimental binding energies [meV] observed for donor impurities
in silicon

P	As	S_b
45.7	53.7	42.7-43.1

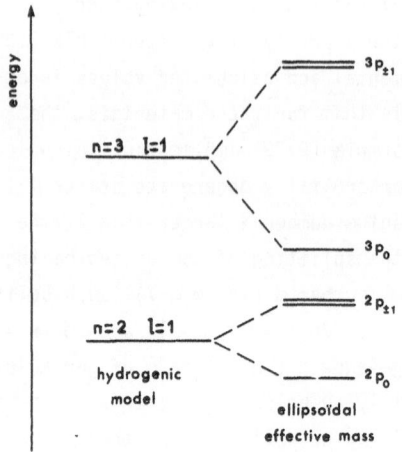

Fig.2.10. Splitting of 2p and 3p states
due to an ellipsoidal effective mass

a large discrepancy which varies with the nature of the impurity. (Such discrepancy is called the chemical shift because it is specific of the impurity involved.)

Table 2.3. Comparison between theoretical and experimental energy spacings (cm^{-1}) for donor impurities in silicon. The experimental values are taken from [2.22], the theoretical ones from [2.23]. (The spacing between excited states, independent of the ground state position, is more adapted than the observed position of the transition from the ground state to do a comparison with the theory since this ground state is not hydrogenic)

Transition	Theory	P	As	S_b
$2p_\pm - 2p_0$	5.11	5.07	5.11	5.12
$3p_0 - 2p_\pm$	0.92	0.93	0.92	0.91
$4p_0 - 2p_\pm$	3.07	3.09	3.10	3.07
$3p_\pm - 2p_\pm$	3.29	3.29	3.28	3.29
$4p_\pm - 2p_\pm$	4.22	4.22	4.21	4.21
$4f_\pm - 2f_\pm$	4.51	4.51	4.49	4.46
$5p_\pm - 2f_\pm$	4.96	4.95	4.94	4.92
$5f_\pm - 2p_\pm$	5.14	5.15	5.14	
$6p_\pm - 2p_\pm$	5.36	5.32	5.32	5.31
$3s - 2p_\pm$	..65			
$3d_0 - 2p_\pm$	2.65	2.64		
$4s - 2p_\pm$	3.55			
$4f_0 - 2p_\pm$	4.07	4.08		
$5p_0 - 2p_\pm$	4.17	4.17		
$5p_0 - 2p_\pm$	4.77	4.76		4.70
$6h_\pm - 2p_\pm$	5.52	5.52		

The reason why the effective mass theory agrees well with experiment for excited states is that these states correspond to large orbits and therefore obey the main assumptions of the theory: the field $U(r)$ is coulombic and varies slowly far from the impurity. However, for the ground state, the amplitude of the wave function in the vicinity of the impurity (in the "central cell") where the field $U(r)$ varies rapidly, is large. Consequently, a correction has to be made inside this central cell. It consists of taking an "effective charge" $Z(r)$ so that

$$U(r) = Z(r) \frac{e^2}{r} \quad . \tag{2.81}$$

Of course, a fit can be obtained by choosing the right amplitude and variation with r for this effective charge Z when the electrons of the inner shells (the core electrons) of the impurity are the same as those of the atoms of the crystal. However, when the impurity has less core electrons than the crystal atoms, the effective potential is repulsive in the central cell and does not give rise to any bound state; for impurities which have more core electrons than the crystal atoms the potential is attractive and gives rise to a deep level. In these cases the effective mass theory with such a correction is not sufficient. It is then combined with a pseudopotential technique. This technique avoids the difficulty of choosing the potential seen by the electron by replacing the Schrödinger equation by a simpler equation not containing the true potential, but whose eigenvalues are the same.

It is possible to obtain some of the states when others are known. This is the situation we are faced with: the "deep" states, which are those of the isolated atoms, are known and we are interested in the "shallow" states in the vicinity of the bands. Separating these deep and shallow states, which we call respectively j and k, we must solve

$$H\psi_k = E_k\psi_k \qquad (2.82)$$

and

$$H\psi_j = E_j\psi_j \quad , \qquad (2.83)$$

where the sets of ψ_j, E_j are known (since they are atomic states). We search for a solution:

$$|\psi_k\rangle = |\phi_k\rangle - \sum_j |\psi_j\rangle \langle\psi_j|\psi_k\rangle \qquad (2.84)$$

(the j and k states are orthogonal).

With such an expression for ψ_k, the Schrödinger equation (2.82) becomes

$$H|\phi_k\rangle - \sum_j H|\psi_j\rangle \langle\psi_j|\psi_k\rangle = E_k|\phi_k\rangle - \sum_j E_k|\psi_j\rangle \langle\psi_j|\psi_k\rangle \quad , \qquad (2.85)$$

which is of the form

$$(T + V_p)|\phi_k\rangle = E_k|\phi_k\rangle \quad , \qquad (2.86)$$

when the pseudopotential V_p replaces the true potential V in H = V + T (T is the kinetic energy term). The expression of this pseudopotential is given by

$$V_p|\phi_k> = V|\phi_k> - \sum_j (E_j|\psi_j> - E_k|\psi_j>) <\psi_j|\psi_k> \qquad (2.87)$$

Equation (2.86), which uses the pseudowave function ϕ_k, has the same eigenvalues as (2.82). The pseudopotential V_p exhibits amplitudes smaller than V and the $|\phi_k>$ vary slowly. The problem is reduced to the choice of this pseudopotential.

In conclusion, the effective mass theory is limited in practice to very few impurities, those which belong to the elements of adjacent columns in the periodic table. Actually the theory does not apply to all of these elements because of the possible lattice distortion and relaxation their presence can induce. It is quite possible that part of the chemical shift could be due to such an effect. Consider, for instance, the case of nitrogen in diamond, which is expected to play the role of a donor. Electron paramagnetic resonance has shown [2.24] that the fifth valence electron of the nitrogen atom remains localized on one of the four bonds, thus inducing a Jahn-Teller distortion [Ref. 1.1, Chap.8]. Such distortion brings the localized level associated with this fifth electron down into the forbidden gap and the resulting level is no longer shallow.

2.4.3 Pairing Effects

Evidence of pairing, i.e., of the presence of donor-acceptor pairs, can be obtained from electronic transport properties, vibrational properties, and radiative recombination.

According to (2.64) or (2.70) pairing decreases the binding energy. Such a decrease has been observed in diamond and cadmium telluride. The measure of the change in the binding energy can give an estimate of the concentration N of the defects, forming long distance pairs, i.e., corresponding to the distribution $g_2(R)$, since the average distance R is related to N through

$$\frac{4}{3}\pi R^3 = \frac{1}{N} \qquad (2.88)$$

(see Fig.2.11). However, electrical measurements are not accurate enough to provide the density of states and therefore the shape of the distribution $g_2(R)$. We shall see [Ref.1.1, Chap.11] that additional information on the existence of pairs can be obtained from the measurement of the free carrier mobility: the pairs act as dipolar scattering centers and consequently give rise to an anomalously high mobility. Figure 2.11 shows the binding energy of a hole (full line) in boron doped diamond versus the cubic root of the impurity concentration N [2.25]. This energy follows only qualitatively the

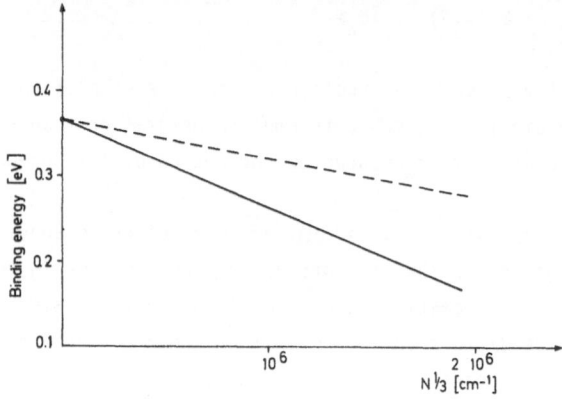

Fig.2.11. Variation of the binding energy of holes in diamond versus the cubic root of the boron concentration [2.26]. The dashed line corresponds to the value of $\alpha(4 \times 10^{-8}$ eV cm) provided by the theory

theoretical law (dashed line)

$$E_B = E_{B0} - \alpha N^{1/3}$$

with $\qquad \alpha = \dfrac{e^2}{\varepsilon} \left(\dfrac{4\pi}{3}\right)^{1/3}$ $\qquad\qquad\qquad\qquad\qquad\qquad$ (2.89)

($E_{B0} = 0.37$ eV, $\alpha = 4 \times 10^{-8}$ eVcm). This can be due to the fact that: a) the pair distribution function is not random, or b) the binding energy is screened by the free carriers [2.26].

Localized vibrational modes (Chap.5) of pairs have been commonly reported. In order to see localized modes associated with an impurity using infrared absorption, it is necessary to cancel the free-carrier concentration, i.e., to compensate the doping impurity. This is done by the diffusion of another impurity of a type opposite to the type of the dopant and pairing occurs naturally during this diffusion. For instance, interstitial lithium, a n-type impurity which can be incorporated at relatively low temperature, is used to compensate boron in silicon, resulting in the formation and observation of lithium-boron (Li-B) pairs. The ratio of the amplitudes of the spectra associated with boron and Li-B pairs provide the percentage of the impurities involved in pairing. The dependence of these amplitudes with the temperature of diffusion can give the pairing energy [2.27]. From the value of this energy, the average distance can be estimated when a Coulomb interaction is assumed (the pairing energy of 0.39 eV found for the Li-B pair indicates a nearest approach distance).

Radiative recombination of electrons on donors with holes on acceptors, has been observed in Group IV as well as in II-VI and III-V semiconductors. Equation (2.71) is found to fit the spectra observed,which sometimes contain terms due to the creation of phonons. A luminescent spectrum is identified as arising from the transfer of an electron from a donor to an acceptor by the temperature dependence of the intensity and also by the shift of the spectrum to higher energies with an increasing density of excitation [2.28]. This shift, due to the saturation of distant pairs compared to close pairs because of the difference in their radiative lifetime (the closer the elements of a pair, the higher the overlap of the corresponding wave functions and the faster the probability of transition), is typical of a donor-acceptor pair transition. The best evidence for radiative recombination of pairs is found in III-V compounds. There the low temperature spectrum consists of scores of lines near the band edge. These lines originate from the recombination of electrons and holes bound to donor-acceptor pairs having different distances [2.29]. Indeed, the distribution function g(R) derived in Sect.2.4 for a continuous system is actually constrained to precise lattice sites and is the sum of delta functions [2.30].

3. Simple Theory of Deep Levels in Semiconductors

This chapter is an introduction to the theory of deep electronic states in semiconductors. We shall use the tight-binding approximation which, in this context, has a great number of merits: a) it can describe the main physical properties of the bulk semiconductor; b) it leads to fairly simple calculations; and c) it gives an essentially correct description of simple defects such as the single vacancy. It is thus ideally suited to an introductory survey of the main electronic properties associated with defects.

The essential characteristics of the defects we treat here is that they introduce electronic energy levels located deep within the gap. The corresponding eigenfunctions decrease exponentially away from the defect and their extension is smaller than for the shallow levels discussed in Chap.2. As shown in this chapter, the effective mass theory is not valid in this case. Usually, the situation is intermediate between a shallow localization and a very strong localization. There exist accurate methods of calculation for both limits but, at the moment, this is not the case for the intermediate situation which we have to deal with and for which specific methods of calculation have to be derived.

This chapter is organized as follows. In the first part we develop elementary applications of the tight-binding method which provide very simple models and are generally found to give an essentially correct description of the physical properties associated with the defects. We recall the principles of the tight-binding approximation and discuss its usefulness and its limits of validity. We then present its application to sp covalently bonded systems as is the case for Group IV semiconductors. We give a direct treatment of the vacancy in a sp bonded linear chain in order to illustrate general properties associated with bound states in semiconductors. We finally describe and discuss the defect molecule model for several simple point defects. In the second part we introduce a more general method of calculation, the Green's function method, which can in principle yield exact results for a given defect perturbative potential. Its application will be done here within the tight-binding context, allowing comparison with the simple models derived in the first part. More elaborate calculations based on Green's function theory will be described in Chap.4, which focuses on many-electron effects, neglected here.

3.1 Elementary Tight-Binding Theory of Defects

This section develops the application of the tight-binding approximation to the deep-level problem. A "realistic" version of this method, which has become quite popular in recent years, is nothing more than a semiempirical technique previously used in chemistry. We can hope that its success will be the object of more theoretical justification in the future, but we shall not discuss that matter here. We prefer to view the tight-binding Hamiltonian (defined below) as a simulation of the true Hamiltonian, the parameters of which are determined semiempirically.

3.1.1 Basic Principles of the Tight-Binding Approximation

Usually one considers that the electrons can be treated independently, the many-electron effects being incorporated in the effective potential energy experienced by each electron (this matter will be discussed further in Chap.4). Another simplification commonly used is to assume that each electron obeys the same individual Schrödinger equation

$$H\psi_k = \varepsilon_k \psi_k \quad , \tag{3.1}$$

where ε_k is the kth energy level, and ψ_k the corresponding eigenfunction.

The basic process of the tight-binding approximation[1] is to expand the wave function ψ_k using a basis formed by the free-atom eigenstates $\chi_{i\alpha}$ ($\chi_{i\alpha}$ is the αth state of atom i)

$$\psi_k = \sum_{i,\alpha} a^k_{i\alpha} \chi_{i\alpha} \quad . \tag{3.2}$$

This is the linear combination of atomic orbitals (LCAO) method. The solutions of (3.1) can be expressed in terms of the matrix elements of the Hamiltonian

$$H_{i\alpha,j\beta} = \langle \chi_{i\alpha} | H | \chi_{j\beta} \rangle \quad . \tag{3.3}$$

---

[1] For a review on the multiple applications of tight-binding see [3.1,2].

In principle, the atomic orbitals centered on different atoms are not ortho-gonal, so that we must consider the interatomic overlap integrals defined by

$$S_{i\alpha,j\beta} = \langle \chi_{i\alpha} | \chi_{j\beta} \rangle \quad . \tag{3.4}$$

The solutions of (3.1) are thus given by the roots of

$$\det |H_{i\alpha,j\beta} - \varepsilon S_{i\alpha,j\beta}| = 0 \quad . \tag{3.5}$$

The major disadvantage of the LCAO method is that, in general, (3.5) contains matrix elements which do not converge rapidly in real space (because the $S_{i\alpha,j\beta}$ are not always small and decrease slowly with interatomic distance) so that its resolution requires heavy numerical computation. A further ap-proximation consists in the neglect of all overlap integrals: this is the tight-binding approximation. Viewed in this way, it is in principle less accu-rate than the LCAO method to which it is only an approximation. However, in practice it has mainly been used in a "semiempirical" or "realistic" manner, the Hamiltonian matrix elements being adjusted to reproduce some known ex-perimental properties (such as the band structure of bulk crystals) and then used to predict other physical properties.

The simplest version of tight binding, in which only nearest neighbors interactions are included, was first originated in quantum chemistry (the simple Hückel theory) for hydrocarbons where it encountered an impressive success. More recently numerous tight-binding investigations have been made in solid state physics, leading to a similar success especially for transi-tion metals and covalent semiconductors.

3.1.2 The Tight-Binding Matrix Elements for sp Covalently Bonded Solids

Consider the case of the elements of Group IV (C, Si, Ge, Sn) which have four valence electrons, the free-atom configuration being s^2p^2. The basic idea of the tight-binding treatment is to take, as basis set for the valence electrons in the solid, the free-atom valence orbitals of each atom, i.e., one s orbital χ_{is} and three p orbitals χ_{ip_x}, χ_{ip_y}, and χ_{ip_z} on each atom i (Ox, Oy and Oz are a fixed set of cartesian coordinates). Using this notation the whole wave function ψ_k of (3.2) is written

$$\psi_k = \sum_i \left(a_{is}^k \chi_{is} + a_{ip_x}^k \chi_{ip_x} + a_{ip_y}^k \chi_{ip_y} + a_{ip_z}^k \chi_{ip_z} \right) \quad . \tag{3.6}$$

The unknown coefficients $a_{i\alpha}^k$ can be determined by diagonalization of the Hamiltonian matrix $H_{i\alpha,j\beta}$. These matrix elements are often restricted to first or second nearest neighbors. They are obtained by a least-square fit to the known band structure (often the valence bands and the lowest conduction band).

While this procedure is systematic, it does not give a good physical insight into the covalent bond problem. It is interesting to make first a basis change (which will not change the final result), so that we can classify the interactions by decreasing order of importance, the leading term being representative of what might be called the perfect covalent limit. This goal is achieved for covalently bonded systems by defining sp hybrid orbitals. Because these hybrids have a strong positive lobe pointing into a given direction, it is clear that only pairs of hybrids whose positive lobes are overlapping will lead to important matrix elements of H.

The construction of the sp hybrids is based on the angular properties of the p orbitals. The angular parts of χ_{ip_x}, χ_{ip_y} and χ_{ip_z} can be written as $\hat{i} \cdot \underline{r}/r$, $\hat{j} \cdot \underline{r}/r$, and $\hat{k} \cdot \underline{r}/r$, where $\hat{i}$, $\hat{j}$, $\hat{k}$ are the unit vectors of the O_y, O_y, and O_z axes. Any orthonormal combination of p orbitals defines another p orbital χ_{ip_α}, of angular part $\hat{\underline{a}} \cdot \underline{r}/r$, whose lobes are thus along the direction defined by the unit vector $\hat{\underline{a}}$ (Fig.3.1).

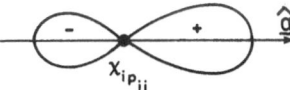

Fig.3.1. Angular dependence of a p orbital pointing in a direction $\hat{\underline{a}}$

Assume now that an atom i has n equivalent bonds with its nearest neighbors j, of unit vectors $\widehat{\underline{ij}}$. One can define n equivalent sp^n hybrids φ_{ij} under the form

$$\varphi_{ij} = \frac{\chi_{is} + \lambda \chi_{ip_{ij}}}{\sqrt{1+\lambda^2}} , \qquad (3.7)$$

where $\chi_{ip_{ij}}$ is the p orbital along the direction $\widehat{\underline{ij}}$. The coefficient λ must be determined so that the different φ_{ij} are orthogonal. This leads to the condition

$$1 + \lambda^2 \cos\theta = 0 , \qquad (3.8)$$

where θ is the angle between two equivalent bonds. For n bonds which are equivalent in space, the sum of unit vectors $\sum_j \widehat{\underline{ij}}$ vanishes. Projecting this identity onto one of these unit vectors $\widehat{ij}$ leads to the fact that $1+(n-1)$ $\cos\theta$ is zero, i.e., $\cos\theta$ is equal to $-1/(n-1)$ and λ to $\sqrt{n-1}$. This applies to fourfold (C, Si, Ge, Sn), threefold (graphite) and twofold coordinations respectively. For such values of λ ($\lambda = \sqrt{3}, \sqrt{2}, 1$) the orbital φ_{ij} has a strong positive lobe along $\widehat{ij}$ and a small negative lobe in the opposite direction (Fig.3.2).

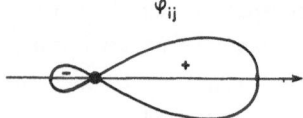

Fig.3.2. Angular dependence of a sp hybrid orbital

For the crystalline form of the Group IV elements, the unit vectors $\widehat{ij}$ are given by $(111)/\sqrt{3}$, $(1\bar{1}\bar{1})/\sqrt{3}$, $(\bar{1},1,\bar{1})/\sqrt{3}$, $(\bar{1},\bar{1},1)/\sqrt{3}$ along the cube axes. The φ_{ij} pointing into these directions can thus be written as

$$\varphi_{i,111} = \frac{1}{2}(x_{is} + x_{ip_x} + x_{ip_y} + x_{ip_z})$$

$$\varphi_{i,1\bar{1}\bar{1}} = \frac{1}{2}(x_{is} + x_{ip_x} - x_{ip_y} - x_{ip_z})$$

$$\varphi_{i,\bar{1}1\bar{1}} = \frac{1}{2}(x_{is} - x_{ip_x} + x_{ip_y} - x_{ip_z})$$

$$\varphi_{i,\bar{1}\bar{1}1} = \frac{1}{2}(x_{is} - x_{ip_x} - x_{ip_y} + x_{ip_z}) \quad . \tag{3.9}$$

We can now examine the matrix elements of H in the basis of these sp^3 orbitals. Let us consider first the intraatomic elements. By symmetry, in the s,p atomic basis, they reduce to two independent terms which are

$$\langle x_{is}|H|x_{is}\rangle = E_s$$

$$\langle x_{ip_\mu}|H|x_{ip_\mu}\rangle = E_p \quad ; \quad \mu = x,y,z \quad . \tag{3.10}$$

These quantities define the s and p atomic levels in the solid, which are generally found to be close to the free atom values (especially for $E_p - E_s$ which is of interest here) so that E_s and E_p are negative and the quantity $E_p - E_s$ is positive. This last quantity is usually termed the "promotion energy", i.e., the energy necessary to promote an electron from the ground s state to the excited p state, or the whole atom from its s^2p^2 ground state to a hypothetical valence sp^3 state which favors the covalent bonding, as we shall see later. Its order of magnitude is 6-8 eV [3.3].

Using the sp^3 basis, it is a simple matter to show that the intraatomic elements are:

$$\langle\varphi_{i\alpha}|H|\varphi_{i\beta}\rangle = \frac{E_s + 3E_p}{4} \qquad \text{for any } \alpha = \beta$$

and

$$\langle\varphi_{i\alpha}|H|\varphi_{i\beta}\rangle = \frac{E_s - E_p}{4} \qquad \text{for } \alpha \neq \beta \qquad (3.11)$$

Note that the energy $\Delta = (E_s - E_p)/4$ is of course negative. The energy $\bar{E} = (E_s + 3E_p)/4$ is the sp^3 average energy; it is of no relevance in band struc-ture problems where only relative energies are needed. We shall therefore take it as the origin of energies.

Let us now investigate the nearest neighbors interactions. There are four independent such interactions when a two-center approximation is used [3.4]. They are pictured in Fig.3.3. The most important one in magnitude is β con-necting two sp^3 orbitals centered on nearest neighbours and pointing towards each other. The three other terms β', β_t and β_c are usually much smaller. The values of these various terms for silicon [3.5] are given in Table 3.1.

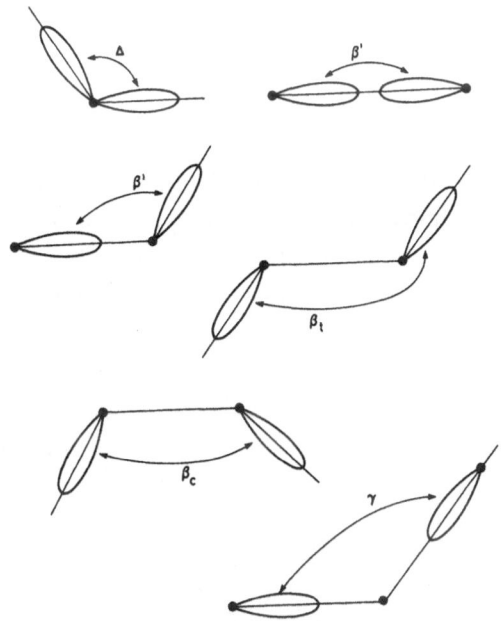

Fig.3.3. Representation of the various interactions between hybrid orbitals belonging to the same atom (Δ), first neighbors ($\beta,\beta',\beta_t,\beta_c$), and second neighbors (γ)

We can go beyond first nearest neighbors. In the vacancy case the term γ, corresponding to second nearest neighbors connecting sp^3 orbitals pointing towards the same nearest neighbor (Fig.3.3), turns out to be of particular importance as will be discussed later. Its value for silicon is also given in Table 3.1.

Table 3.1. Tight-binding parameters for silicon [eV]

β	Δ	β'	β_c	β_t	γ
-3.75	-1.12	-0.51	-0.33	+0.22	-0.25

3.1.3 Formation of the Band Structure

We shall consider the formation of the band structure in successive steps, which allows a very simple interpretation of the nature of the covalent bond. For this we note that in the covalent systems of interest here, β is the leading term, so that the zeroth-order approximation will be obtained by taking $\beta \neq 0$ and neglecting all other interactions. This leads to the "molecular" or "bond orbital model".

a) The Molecular Model

When we consider only β interactions, the problem is very simple. The sp^3 orbitals are connected by pairs. For instance, the orbital φ_{ij} pointing from atom i towards j interacts only with φ_{ji}. For each pair of orbitals the secular problem reduces to the diagonalization of a 2×2 matrix whose elements are

$$\langle \varphi_{ij} | H | \varphi_{ij} \rangle = \langle \varphi_{ji} | H | \varphi_{ji} \rangle = 0$$

$$\langle \varphi_{ij} | H | \varphi_{ji} \rangle = \beta \quad , \tag{3.12}$$

i.e.,

$$\begin{bmatrix} 0 & \beta \\ \beta & 0 \end{bmatrix} \quad . \tag{3.13}$$

The eigenvalues and eigenfunctions of this matrix are

$$E_{B \atop A} = \pm \beta$$

$$\Psi_{B \atop A, ij} = \frac{\varphi_{ij} \pm \varphi_{ji}}{\sqrt{2}} . \qquad (3.14)$$

The indices B or A refer to the bonding (+) and antibonding (-) states, respectively, similar to the hydrogen molecule to which this problem is formally equivalent. Thus, in this limit, we can view the whole solid as a collection of diatomic molecules whose number is equal to the number 2N of bonds between nearest neighbors. The band structure consists of two levels E_B and E_A, having a degeneracy equal to 2N, the lowest one E_B being completely filled (the total number of electrons is 4N). This two-levels scheme describes the essential step in the formation of the valence and conduction bands of covalent systems. In this model, the charge density is a superposition of the individual bond-charge densities, a fact which is quantitatively confirmed by more elaborate calculations [3.6]. Many other physical properties have been interpreted on the basis of this "molecular" description [3.7] whose simplicity and accuracy is thus very appealing. It has also been generalized to III-V and II-VI compounds, as well as to other covalent systems, with similar success.

b) Broadening Effects. A Simple Description

If we include further interactions, the molecular model will no longer remain valid. This will cause a broadening of the bonding and of the antibonding levels as well as a coupling between them. We shall first use a simple model whose advantage is that it has an analytical solution. For this, we only add the interaction Δ corresponding to the $s \rightarrow p$ atomic promotion energy. Such a model has been widely used [3.8] because its simplicity allows its extension to amorphous systems [3.9]. We thus write any eigenstate Ψ under the form

$$\Psi = \sum_{i,j} a_{ij} \varphi_{ij} , \qquad (3.15)$$

the sum running over pairs of nearest neighbors i and j (the two possibilities ij and ji have to be considered separately since the sp^3 orbital φ_{ij} pointing from i to j differs from φ_{ji}).

Projecting the Schrödinger equation onto the two coupled basis states φ_{ij} and φ_{ji} we get the two equations

$$Ea_{ij} = \Delta \sum_{j' \neq j} a_{ij'} + \beta a_{ji}$$

$$Ea_{ji} = \Delta \sum_{i' \neq i} a_{ji'} + \beta a_{ij} \quad , \tag{3.16}$$

valid for any pair of atoms i and j. Equations (3.16) can be transformed into

$$(E + \Delta)a_{ij} = \Delta S_i + \beta a_{ji}$$

$$(E + \Delta)a_{ji} = \Delta S_j + \beta a_{ij} \quad , \tag{3.17}$$

where S_i is the sum $\sum_j a_{ij}$ of the coefficients connecting atom i to its four nearest neighbors. Equations (3.17) can be transformed by replacing a_{ji} in the first one by its value deduced from the second. We then obtain

$$[(E + \Delta)^2 - \beta^2]a_{ij} = \Delta(E + \Delta)S_i + \beta \Delta S_j$$

$$[(E + \Delta)^2 - \beta^2]a_{ji} = \Delta(E + \Delta)S_j + \beta \Delta S_i \quad , \tag{3.18}$$

which can be summed over the four nearest neighbors to give for any atom i

$$\varepsilon S_i = \sum_j S_j \quad , \quad \text{with}$$

$$\varepsilon = \frac{(E - \Delta)^2 - (\beta^2 + 4\Delta^2)}{\beta \Delta} \quad . \tag{3.19}$$

Equations (3.19) reduce the band structure problem to a one-band problem where ε is the energy and where the interactions between nearest neighbors are equal to unity. Topological theorems [3.10] show that ε lies in the interval [-4, +4], so that we obtain two bands given by

$$E = \Delta \pm \left(\beta^2 + 4\Delta^2 + \beta \Delta \varepsilon \right)^{\frac{1}{2}} \quad . \tag{3.20}$$

These two bands correspond to $S_i \neq 0$, but there are also nontrivial solutions for $S_i = 0$. These can occur when [see (3.18)]

$$E = -\Delta \pm \beta \quad , \tag{3.21}$$

which is the equation of two flat bands.

These results depend only on the structure of the Hamiltonian and not on whether the system is crystalline or not. The energy E can be plotted against ε (Fig.3.4), which gives the band structure as compared to the two levels of the molecular model. The conclusions which emerge are the following.
a) The molecular levels are broadened by an amount $4|\Delta|$.

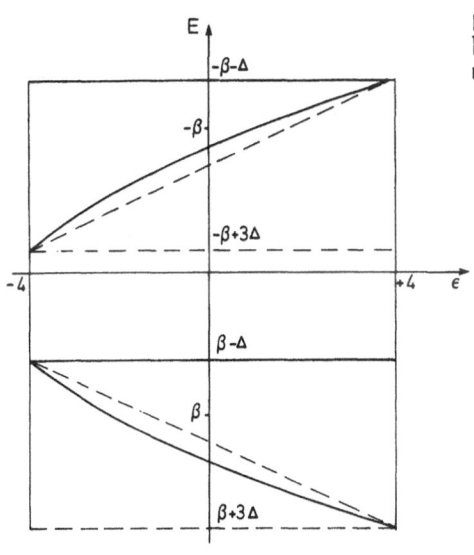

Fig.3.4. Band structure resulting from broadening effects on the molecular model

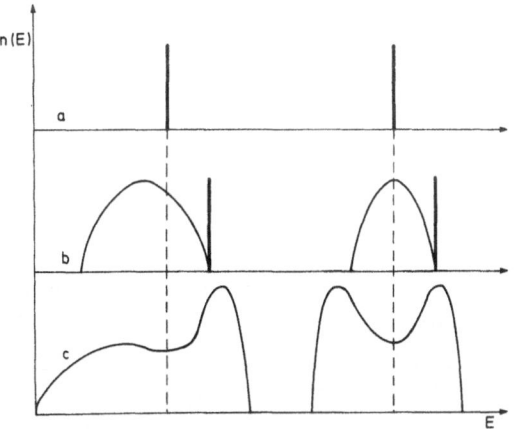

Fig.3.5a-c. Density of states in the molecular model (a), when broadening effects are taken into account (b), and in a more realistic case (c)

b) The covalent situation occurs as long as there is no band crossing, i.e., $|\beta| > 4|\Delta|$.

c) The barycenter of the valence band is lower than β due to the interaction between the bonding and antibonding states.

The schematic density of states in this model is pictured in Fig.3.5b where it is compared to the two delta functions density of states of the molecular model (Fig.3.5a) and to a more refined version (Fig.3.5c).

c) Refinements

The inclusion of further interaction such as β', β_c, β_t, γ will not alter the above qualitative conclusions for the valence band except that they broaden it further (especially the delta function peak). They are however essential in giving the fine details of the forbidden gap and the bottom of the conduction band .

3.1.4 The Vacancy in a Covalently Bonded Linear Chain

We introduce this model as an analytical illustration of what occurs in a three-dimensional system. We consider a linear chain of identical atoms with two valence electrons per atom, taking as basis one s and one p orbital oriented along the chain. We will not consider the electrons in p orbitals perpendicular to the chain that do not participate to the covalent bonds of interest. We can use the same arguments as in the two previous sections and build sp hybrids $\varphi_{i,i+1}$ and $\varphi_{i,i-1}$ pointing from atom i towards its two neighbors $i+1$ and $i-1$ (Fig.3.6). They are

$$\varphi_{i,i\pm1} = \frac{\varphi_{is} \pm \varphi_{ipx}}{\sqrt{2}} \quad . \tag{3.22}$$

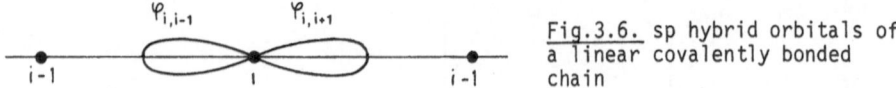

$\varphi_{i,i-1}$ $\varphi_{i,i+1}$

i-1   i   i-1

Fig.3.6. sp hybrid orbitals of a linear covalently bonded chain

As for the model used in Sect.3.1.3b, we consider the nearest neighbors' interaction β of hybrids engaged in the same bond and Δ, equal here to $(E_s - E_p)/2$. Similar arguments as those of this section, with $-2 \leq \varepsilon \leq +2$, lead to the density of states for the molecular model (Fig.3.7a) and for the complete band structure (Fig.3.7b). We have to calculate the perturbation introduced by the creation of a vacancy in such a chain, which we consider to be produced by the removal of the atom located at the site i = 0 of the chain. In view of the restricted set of interactions which we have considered, the problem is equivalent to two independent identical semiinfinite chains of Fig.3.8. We shall therefore treat only one of these chains with i > 0 and assign to the corresponding levels a degeneracy of 2.

Let us first begin with the molecular model, i.e., the limit $\Delta/\beta \to 0$. In this case, any coupled set of sp orbitals will lead to its bonding and anti-bonding state at energies β and $-\beta$. Only φ_{10} will remain uncoupled and thus will be an eigenstate at its atomic energy E = 0 (the average sp state

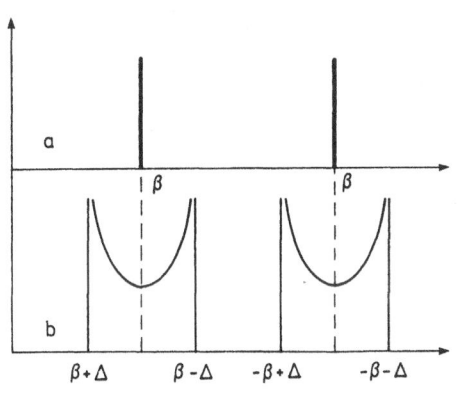

Fig.3.7a,b. Density of states for a
linear chain in the molecular model
(a) and when broadening effects are
taken into account (b)

Fig.3.8. Representation of a vacancy in a linear chain

$(E_s + E_p)/2$ is taken as the origin of the energies). The density of states
for the system perturbed by the vacancy is shown in Fig.3.9. The level at
$E = 0$ is a localized state at the vacancy. It is twofold degenerate and its
eigenstates are φ_{10} and φ_{-10}. These states are usually called "dangling orbitals" because they do not form bonds as in the perfect crystal.

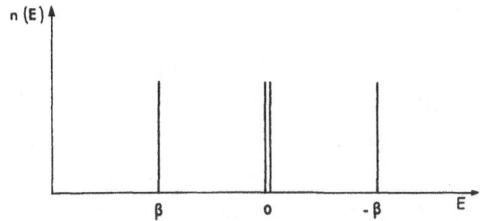

Fig.3.9. Density of states of a
linear chain containing a vacancy
in the molecular model

What happens to these states when we include the interaction Δ? As before,
the bonding and antibonding levels will broaden into valence and conduction
bands. However, here we are mainly interested to what happens to the localized
states at $E = 0$. The interaction of a dangling orbital such as φ_{10} with its
neighboring bonding and antibonding states is identical and equal to $\Delta/\sqrt{2}$. The
problem remains completely symmetric in energy so that $E = 0$ will remain a
solution. Let us then look at the wave function of this "deep level" within
the gap. Its coefficients are given, in general, by the following set of
equations:

$$\begin{cases} Ea_{10} = \Delta a_{12} \\ Ea_{12} = \Delta a_{10} + \beta a_{21} \end{cases} \qquad i < 2$$

and

$$\begin{cases} Ea_{i,i-1} = \Delta a_{i,i+1} + \beta a_{i-1,i} \\ Ea_{i,i+1} = \Delta a_{i,i-1} + \beta a_{i+1,i} \end{cases} \qquad i \geq 2 \qquad (3.23)$$

For $E = 0$, we find that all $a_{i,i+1}$ vanish while all $a_{i,i-1}$ are given by

$$a_{i+1,i} = - \frac{\Delta}{\beta} a_{i,i-1} = \left(-\frac{\Delta}{\beta}\right)^{i} a_{10} \qquad . \qquad (3.24)$$

The deep-level wave function is thus exponentially decreasing and its normalized expression is given by:

$$\Psi = \sqrt{1-\Delta^{2}/\beta^{2}} \sum_{i=0}^{\infty} \left(- \frac{\Delta}{\beta}\right)^{i} \varphi_{i+1,i} \qquad . \qquad (3.25)$$

The condition for the convergence of this expression is $\Delta/\beta < 1$. The contribution of the dangling orbital $|<\varphi_{10}|\Psi>|^{2}$ to the wave function of the bound state is $1 - (\Delta^{2}/\beta^{2})$, which tends to unity when the ratio of the width of each individual band $2|\Delta|$ to the average gap $2|\beta|$ tends to zero, i.e., in the perfect covalent limit.

3.1.5 The Vacancy in a Covalent Crystal

We consider here one of the basic point defects for which there is a wealth of experimental results (there exists detailed information [3.11] in the case of silicon). The local situation is depicted in Fig.3.10: the vacancy corresponds to the removal of atom 0. On this figure are also drawn the sp^{3} orbitals labelled φ_{i0} ($i = 1$ to 4 in our notation) which point towards the vacancy site. The simplest analysis of the situation can again be made on the basis of the molecular model. There are now four dangling orbitals at energy $E = 0$ [the origin of energy being now $(E_{s} + 3E_{p})/4$] while all other sp^{3} states are engaged in bonds at energies $\pm \beta$. We thus obtain a fourfold degenerate bound state, completely localized on the four dangling orbitals which point towards the vacancy site.

To go beyond this simple description we first introduce the interaction Δ alone, as in Sect.3.1.3b. In this case it can be shown that for a Bethe lattice (i.e., a fourfold coordinated lattice with no closed loops of atoms, formally

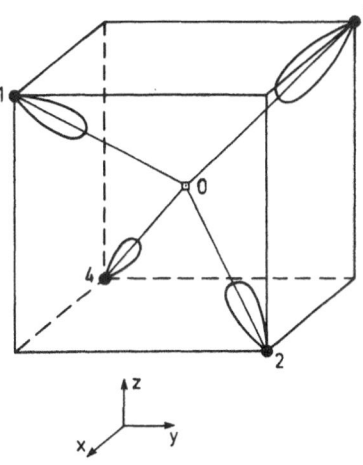

Fig.3.10. Molecular model of a vacancy in a Group IV semiconductor

equivalent to a linear chain) $E = 0$ remains an exact fourfold degenerate localized state, the contribution of the dangling orbitals to the bound-state wave function now being $1 - (3\Delta^2/\beta^2)$. This result remains practically unaltered for the diamond lattice. From what has just been said, it is clear that a nearest-neighbors' tight-binding model is not able to describe correctly any possible splitting of the localized levels of the vacancy. This is due to the fact that the model does not include the direct interaction γ between pairs of dangling orbitals (see Fig.3.3); any meaningful model must take this inter-action into account. We therefore now introduce it and analyze qualitatively what will then happen. (More quantitative results will be given in the second part of this chapter.) For this, we group the tight-binding interactions into three classes:

1) interactions within the subset of the four dangling orbitals φ_{i0} (i.e., γ),
2) interactions between the sets of bonding and antibonding orbitals of the molecular model,
3) coupling between the two subsets.

The effect of the first class can be discussed readily. It leads to what is called "the defect molecule model of the vacancy", which has provided an extremely useful basis for the interpretation of experimental data. Within the 4×4 subspace of the dangling orbitals φ_{i0}, we have to diagonalize the matrix (with $\gamma < 0$)

$$\gamma \begin{bmatrix} 0 & 1 & 1 & 1 \\ 1 & 0 & 1 & 1 \\ 1 & 1 & 0 & 1 \\ 1 & 1 & 1 & 0 \end{bmatrix}$$

This can be done directly, or using symmetry arguments. However, this matrix is identical to the one which connects the four sp^3 hybrids φ_{0i} of atom 0 (once γ is replaced by Δ). Its eigenstates are thus expressed in terms of φ_{i0}, exactly as x_{0s}, x_{0p_x}, x_{0p_y}, and x_{0p_z} were expressed in terms of the φ_{0i}. For the situation of Fig.3.10, we have

$$\varphi_{01} = (x_{0s} + x_{0p_x} - x_{0p_y} + x_{0p_z})/2$$
$$\varphi_{02} = (x_{0s} + x_{0p_x} + x_{0p_y} - x_{0p_z})/2$$
$$\varphi_{03} = (x_{0s} - x_{0p_x} + x_{0p_y} + x_{0p_z})/2$$
$$\varphi_{04} = (x_{0s} - x_{0p_x} - x_{0p_y} - x_{0p_z})/2 \quad . \tag{3.26}$$

Inverting this set of equations, we obtain x_{0s}, x_{0p_x}, x_{0p_y}, and x_{0p_z} in terms of φ_{0i}. The eigenfunctions are thus obtained by replacing φ_{0i} by φ_{i0} in these expressions, leading to

$$\left. \begin{aligned} v &= (\varphi_{10} + \varphi_{20} + \varphi_{30} + \varphi_{40})/2 \quad A_1 \\ t_x &= (\varphi_{10} + \varphi_{20} - \varphi_{30} - \varphi_{40})/2 \\ t_y &= (-\varphi_{10} + \varphi_{20} + \varphi_{30} - \varphi_{40})/2 \\ t_z &= (\varphi_{10} - \varphi_{20} + \varphi_{30} - \varphi_{40})/2 \end{aligned} \right\} T_2 \quad . \tag{3.27}$$

As is x_{0s}, v is completely symmetric under the symmetry operations of the tetrahedral point group T_d. It belongs to the irreducible representation A_1 of that group. t_x, t_y and t_z behave as x_{0p_x}, x_{0p_y}, and x_{0p_z} and belong to T_2. As for s and p states, we obtain a lower nondegenerate level at 3γ and a threefold degenerate one at $-\gamma$ (Fig.3.11).

The second class of interactions connect the bonding and antibonding orbitals. Their effect is then to broaden the corresponding levels into valence and conduction bands as we have described previously.

The third type of interactions will finally couple the eigenstates of the defect molecule (3.27), which are purely built from dangling orbitals, with the remaining crystal. It is this class of terms which allow for the delocalization of the bound-state wave function. We now describe what happens in a

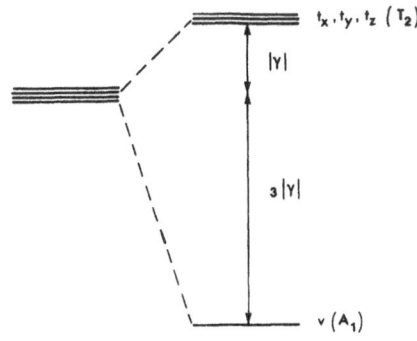
$t_x, t_y, t_z (T_2)$

$|\gamma|$

$3|\gamma|$

$v (A_1)$

Fig.3.11. Fig.3.11. Splitting of the levels associated with the vacancy in the molecular model due to the interaction γ

more explicit manner. For this let us write the final wave function of the localized state Ψ under the form

$$\Psi = a_d \chi_d + \sum_\alpha c_\alpha \varphi_\alpha \quad , \qquad (3.28)$$

where χ_d is one of the eigenstates of the defect molecule and φ_α are the eigenstates of the remaining crystal, a_d and c_α being unknown coefficients. As a consequence of symmetry this can be done for each χ_d occuring in (3.27) in a completely separated manner. Using (3.28) we can then write the equations

$$(E - E_d)a_d = \sum_\alpha V_{d\alpha} c_\alpha$$

$$\qquad (3.29)$$

$$(E - E_\alpha)c_\alpha = V_{\alpha d} a_d \quad ,$$

where E_d and E_α are the eigenvalues of the separate subsystems, and V is the matrix containing the third class of interactions, i.e., coupling the defect molecule to its environment. Injecting the second equation into the first one, we get the equation for the possible energy levels of the coupled system

$$E = E_d + \sum_\alpha \frac{|V_{d\alpha}|^2}{E-E_\alpha} \quad . \qquad (3.30)$$

To discuss the kind of solutions we might obtain, it is important to recall that all E_α are grouped into a valence band and a conduction band. In cases where (3.30) has solutions within the forbidden gap we have a localized state and, if this solution is not far from the corresponding value of E_d, the defect molecule can be considered as a meaningful starting point. As will be shown in the second part of the chapter, this is what is found for the T_2 state, but not for the A_1 state, which falls within the valence band.

Other information is obtained from an analysis of the delocalization of the bound-state wave function. For this, we use the second equation in (3.29) expressing the c_α in terms of a_d. After normalization of Ψ, we obtain

$$|a_d|^2 = \frac{1}{1 + \sum_\alpha \frac{|V_{\alpha d}|^2}{|E - E_\alpha|^2}} \qquad . \qquad (3.31)$$

This shows quite generally that, just as in the simple case of the covalent linear chain, the bound-state wave function (if it exists) is now delocalized, so that $|a_d|^2$ becomes smaller than unity. This delocalization increases with the strength of the average coupling $|V_{\alpha d}|$ and also when the average gap decreases, in which case the average $|E - E_\alpha|$ is decreasing. The defect molecule concept becomes useful only if this delocalization is not too important.

This discussion will be generalized in the second part of this chapter when we shall describe the Green's function formalism applied to the defect problem. We shall also discuss the numerical results corresponding to different models of the vacancy.

3.1.6 The Interstitial

It is interesting to discuss the interstitial using the same models as for the vacancy, because this reveals striking differences between the two defects. We shall show, in particular, that the "defect molecule" model is much less adapted to this case than to the vacancy. Also, only the molecular model will be developed for this situation, which even for this simple description is completely different.

We consider one possible configuration, namely, the tetrahedral interstitial pictured in Fig.3.12. This figure has been drawn with the four sp^3 hybrids φ_{0i} pointing from this atom towards its nearest neighbors. These neighbors have four other nearest neighbors shown in Fig.3.12 for atom 1. In the molecular model of the perfect crystal, atom 1 had its four sp^3 hybrids φ_{1j} (pictured in Fig.3.12 with $j = 1',2',3',4'$) engaged in bonds with the φ_{j1}; the situation is the same for the three other neighbors $i = 2,3$, or 4 of the interstitial. In the molecular model, we can say that each of the φ_{0i} belonging to the interstitial will be coupled to another sp^3 hybrid φ_{i0} pointing from this atom i to atom 0. However, these new hybrids φ_{i0} are obtained from the normal sp^3 hybrids of the perfect crystal φ_{ij} (with $j \neq 0$) by a basis change and they are not orthogonal to them. Thus, the four φ_{0i} of the interstitial are coupled to the normal sp^3 hybrids of its neighbors φ_{ij}.

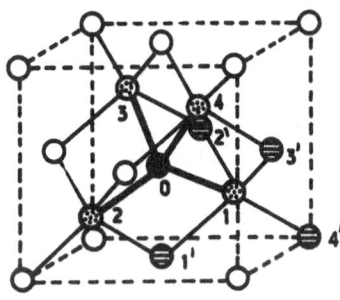

Fig.3.12. The tetrahedral interstitial configuration (0) showing first neighbors
(1 to 4) and second neighbors (1' to 4' for
atom 1)

The interstitial problem in the molecular model requires the diagonalization of four identical 9×9 matrices (one of them corresponding to the basis
φ_{01}, the four φ_{1j}, and the four φ_{j1} with $j = 1',2',3',4'$) all remaining eigenstates being, as usual, bonds and antibonds. To solve this nontrivial problem
in the most economical way, we consider the coupling of φ_{01} with the group of
orbitals φ_{1j} exactly as was done in the set of equations (3.28), identifying
φ_{01} with χ_d. We thus have to determine the eigenstates of the perfect crystal
which are coupled to χ_d. In the molecular model of the perfect crystal, they
are, of course, the bonding and antibonding states,

$$\psi_{\substack{B \\ A}1,j} = \frac{\varphi_{1j} \pm \varphi_{j1}}{\sqrt{2}} \qquad j = 1',2',3',4' \quad , \tag{3.32}$$

of energies $\pm\beta$. To derive (3.30), we now have to evaluate

$$\sum_\alpha \frac{|V_{d\alpha}|^2}{E-E_\alpha} = \sum_j \left(\frac{|<\psi_{B1j}|\varphi_{01}>|^2}{E-\beta} + \frac{|<\psi_{A1j}|\varphi_{01}>|^2}{E+\beta} \right) \quad . \tag{3.33}$$

Here, φ_{01} is only coupled to the φ_{1j} so that (3.33) can be rewritten directly
under the form

$$\sum_\alpha \frac{|V_{d\alpha}|^2}{E-E_\alpha} = \frac{E}{E^2-\beta^2} <\varphi_{01}| \sum_j |\varphi_{1j}> <\varphi_{1j}| |\varphi_{10}> \quad . \tag{3.34}$$

The quantity $\sum_j |\varphi_{1j}><\varphi_{1j}|$ is invariant under any local basis change on atom
1. As a new basis, we choose first the sp^3 hybrid φ_{10} pointing towards the
interstitial and three other independent combinations of φ_{11}, φ_{12}, φ_{13}, and
φ_{14}. As $<\varphi_{01}|\varphi_{10}>$ is equal to β, we easily find that (3.34) can be rewritten
as $(E\beta^2)/(E^2 - \beta^2)$. The eigenvalues are now given by (3.30) with $E_d = 0$,
leading to

$$E[1 - \beta^2/(E^2 - \beta^2)] = 0 \quad . \tag{3.35}$$

The solutions are $E = 0$ and $E = \pm \beta \sqrt{2}$. They are fourfold degenerate because the same solution would hold for $\varphi_{02}, \varphi_{03}, \varphi_{04}$. All other eigenvalues are the simple bonding and antibonding states at $\pm \beta$. The level scheme has thus some similarity (in the molecular model) with the vacancy one, since we find one fourfold degenerate state at $E = 0$. New features arise corresponding to one fourfold state at $E = \beta \sqrt{2}$ (under the valence band level at $E = \beta$) and one fourfold state at $E = -\beta \sqrt{2}$ (above the conduction band level $E = -\beta$).

It is of interest to analyze the wave function of these new states characteristic of the interstitial. For this we simply evaluate $|a_d|^2$, given by (3.31), in the same fashion as was done to obtain (3.34). This leads to

$$|a_d|^2 = \frac{1}{1 + \frac{\beta^2}{2}\left(\frac{1}{(E+\beta)^2} + \frac{1}{(E-\beta)^2}\right)} \quad . \tag{3.36}$$

For $E = 0$, $|a_d|^2 = 1/2$, and the wave function of this state is only 50% localized on the interstitial, in contrast to the vacancy where there is 100% localization in the molecular model. The other states at $E = \pm \beta \sqrt{2}$ only have a 25% localization. Thus, any reasonable defect molecule model for the interstitial should include this effect (this is not the case in the literature [3.12]).

As for the vacancy, further interactions will split each of the fourfold degenerate levels at $E = 0$ and $E = \pm \beta \sqrt{2}$ into a lower nondegenerate component A_1 and a higher threefold degenerate one, T_2. The corresponding splitting must be important because it contains an intraatomic contribution in the interstitial atom itself (the term Δ corresponding to the promotion energy). The $E = 0$ localized states tend to disappear, the A_1 component falling into the valence band, the T_2 component into the conduction band. As for the $E = \sqrt{2}\beta$ states, the A_1 component, which is lowest, can possibly become a state localized under the valence band; the T_2 states being raised are likely to fall within the valence band. The reverse situation holds for the $E = -\sqrt{2}\beta$ states, A_1 being lowered into the conduction band, T_2 being possibly a localized state above the conduction band.

3.1.7 The Substitutional Impurity

Here, we shall consider deep-level impurities for which the short-range potential has the most pronounced effect, so that the effective mass approximation is of no help.

We apply our simple tight-binding model to this situation to see what are
the most important parameters. In a tight-binding framework, the impurity
will be characterized by new matrix elements of the Hamiltonian which differ
from the perfect crystal values only in the neighborhood of the impurity
site.

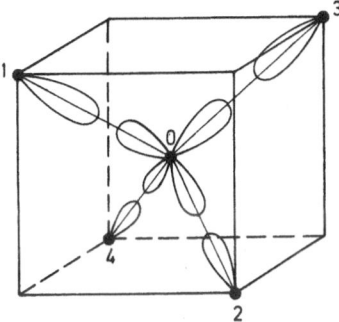

Fig.3.13. Molecular model of a substitutional impurity (o) in a Group IV semiconductor

The first new elements will be those on the impurity site. For an sp
impurity there are new values for the s and p energies at the impurity. In
terms of sp^3 hybrids, this corresponds to a shift of the sp^3 average energy
equal to some value U and a new value of the sp promotion energy which will
be Δ'. We must also consider changes in the interatomic parameters, of which
the most important one is β, taken to be β_1 for sp^3 hybrids connecting the
impurity atom 0 and its nearest neighbors i = 1,2,3, and 4. As usual, let
us begin with the molecular model. The parameters are β for sp^3 hybrids
engaged in bonds between host atoms, β_1 for those engaged in bonds between 0
and its four neighbors, and finally U, the shift in sp^3 energy at the impurity
site. The level structure is again very simple. It first consists of unperturbed bonding and antibonding states at $\pm \beta$, except for the four bonds 0 - i.
For these states we have four identical molecules which are heteropolar.
There are thus four identical 2 × 2 matrices, in the basis of each pair of
sp^3 hybrids φ_{0i}, φ_{i0}, i = 1,4:

$$\begin{bmatrix} U & \beta_1 \\ \beta_1 & 0 \end{bmatrix} ,$$

(3.37)

whose eigenvalues are

$$E_{1\,{}^A_B} = \frac{U}{2} \pm \left[\left(\frac{U}{2}\right)^2 + \beta_1^2\right]^{\frac{1}{2}} . \tag{3.38}$$

In many cases, β_1 is not too different from β and the most important part of the perturbation is due to U. The energies E_{1B}, in the case $\beta_1 = \beta$, are plotted in Fig.3.14 versus the parameter U, in units of $|\beta|$. To derive some conclusions, it is necessary to fix some values of U. The quantity U is equal to the difference in sp^3 energies between the impurity and the host atom.

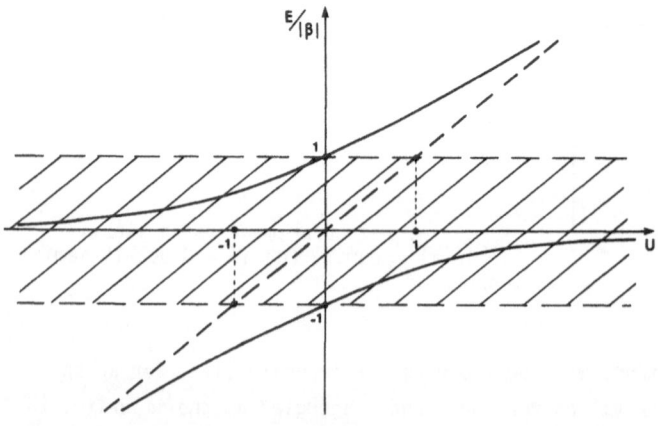

Fig.3.14. Variation of the bonding and antibonding energies, in unit of β, associated with the substitutional impurity versus the shift in sp^3 energy between the impurity and the host atoms

It has been shown [3.13] that a correct order of magnitude for such difference can be deduced from free atom spectra. Such values for the sp^3 energies, obtained from [3.3], are listed in Table 3.2. We can see that for Si and Ge, U becomes negative for impurities of columns V and VI which have more electrons on their outer sp shell and therefore behave as simple or multiple donors. In this case four antibonding molecular states are lowered into the average gap. They give a fourfold degenerate state whose population is equal to the excess number of electrons. The symmetric situation holds also for atoms on the left of column IV for which U is mainly positive. Four bonding states are raised whose population is 8 electrons minus the number of missing electrons, i.e., is, in terms of holes, equal to the number of missing electrons.

We now have to investigate what are the main changes introduced by refinements to this molecular model. We can proceed exactly as for the vacancy, first

Table 3.2. sp^3 energies (eV) for various atoms

	B	C	N	O
	9.7	12.9	16	20.3
	Al	Si	P	Se
	7.3	9.5	11.9	13.4
Zn	Ga	Ge	As	Te
6	7.6	9.5	11.3	12
Cd	In	Sn	Sb	S
5.8	7.3	8.3	10.8	14

considering the four bonding and antibonding states corresponding to the bonds 0 - i as forming a defect molecule, and then coupling it to the remaining crystal. The difference is that there are eight independent orbitals instead of four in the vacancy case.

There are three distinct interactions: V_{BB}, the interaction between two bonding orbitals $\psi_{B,0i}$ and $\psi_{B,0j}$; V_{AA}, the interaction between two antibonding orbitals $\psi_{A,0i}$ and $\psi_{A,0j}$; and finally V_{BA}, the interaction between $\psi_{B,0i}$ and $\psi_{A,0j}$. Again, we can easily diagonalize the 8×8 matrix with the help of symmetry. As for the vacancy case, we build the linear combinations of atomic orbitals

$$
\left.
\begin{aligned}
v_B &= \frac{\psi_{B01} + \psi_{B02} + \psi_{B03} + \psi_{B04}}{2} \\
v_A &= \frac{\psi_{A01} + \psi_{A02} + \psi_{A03} + \psi_{A04}}{2}
\end{aligned}
\right\} \quad A_1
$$

$$
\left.
\begin{aligned}
t_{xB} &= \frac{\psi_{B01} + \psi_{B02} - \psi_{B03} - \psi_{B04}}{2} \\
t_{xA} &= \frac{\psi_{A01} + \psi_{A02} - \psi_{A03} - \psi_{A04}}{2} \\
\vdots \quad &\text{------------------}
\end{aligned}
\right\} \quad T_2 \quad , \tag{3.39}
$$

with similar combinations for ty_B, ty_A, tz_B, tz_A as for the vacancy. Within this basis, the 8×8 matrix becomes factorized into the following form

$$
\begin{bmatrix}
\begin{matrix} E_{1B} + 3V_{BB} & 3V_{BA} \\ 3V_{BA} & E_{1A} + 3V_{AA} \end{matrix} & 0 & 0 & 0 \\
0 & \begin{matrix} E_{1B} - V_{BB} & -V_{BA} \\ -V_{BA} & E_{1A} - V_{AA} \end{matrix} & 0 & 0 \\
0 & 0 & \begin{matrix} E_{1B} - V_{BB} & -V_{BA} \\ -V_{BA} & E_{1A} - V_{AA} \end{matrix} & 0 \\
0 & 0 & 0 & \begin{matrix} E_{1B} - V_{BB} & -V_{BA} \\ -V_{BA} & E_{1A} - V_{AA} \end{matrix}
\end{bmatrix}
$$

$$(3.40)$$

The resulting level scheme is given in Fig.3.15 where we have separated the effect of V_{BB}, V_{AA}, and of V_{BA}, the latter being usually much less important.

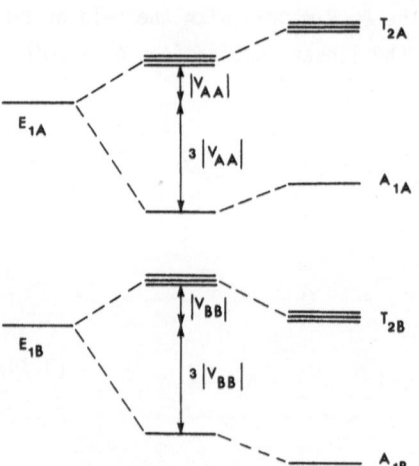

Fig.3.15. Effect of the interactions between bonding and antibonding orbitals of the defect molecule on the defect molecule levels

If we now couple this defect molecule to the remaining crystal, the qualitative consequences will be the same as for the simple vacancy, i.e., only part of these states will be bound states and consequently will be delocalized. Furthermore, it is also useful to notice that the $U \to \pm \infty$ limit gives in both cases the vacancy results because the atomic states of the impurity will be infinitely far in energy from the host crystal levels so that they no longer interact with them. For $U \to -\infty$, the A_{1A} and T_{1A} levels will thus tend towards the A_1 and T_2 states of the vacancy, while for $U \to +\infty$, it is the A_{1B}, T_{2B} states which will tend towards this limit (as we shall see later, the T_2 vacancy state lies within the gap and the A_1 state within the valence band).

We can now describe qualitatively what can happen depending on the sign of U.

a) $U < 0$: For very small U, there should be no bound states. As $|U|$ increases, a A_{1A} bound state will apppear for a given critical value U_1. Then, this bound state will cross the gap and fall within the valence band for another value U_1'. During this increase, a T_{2A} bound state will appear for a third critical value U_2. We cannot infer if U_1' is greater or smaller than U_2 from our qualitative considerations.

b) $U > 0$: The situation is not symmetrical because in this case only the T_{2B} state will be raised within the gap while the A_{1B} will remain within the valence band.

An important conclusion of this simple description of substitutional impurities is that for a strong attractive or repulsive potential U, the impurity levels are pinned to energies corresponding to the vacancy levels. In silicon this corresponds to the T_2 level in the fundamental energy gap. These conclusions can be extended to zinc blende compounds, as was done in [3.14], practically without modification. In the case of GaP, for instance, there will be two sets of levels A_1 and T_2 for the Ga and P vacancies, the P vacancy levels being lower in energy. In [3.14] the A_1 levels corresponding to the impurities N and O at the anion site are calculated in the alloys $GaAs_{1-x}P_x$ as a function of the alloy composition x. It is convincingly shown that the A_1 level does not follow the conduction band when x is varied, but instead follows the anion vacancy level (which falls into the gap for these compounds). The reason is that the perturbation U is strongly attractive (-4 to -8 eV, from Table 3.2) so that we are not far from the vacany situation.

3.2 Green's Function Theory of Defects: Tight-Binding Application

We detail here an exact method of calculation particularly adapted to the
deep-level problem, namely, the Green's function technique.[2] We first present
general arguments, valid for any sort of basis states, then make specific
applications to the tight-binding approximation. Possible refinements will
be left for the next chapter which deals with many-electron effects. We also
discuss different versions of the Green's functions techniques, for instance,
methods of moments, which have been popular recently for transition metals
as well as for covalent systems.

3.2.1 Relation between the Resolvant or Green's Operator and the Density of States

A central quantity in the study of the electronic structure of solid state
systems is the density of states $n(E)$, defined as the number of states per
unit energy lying in the energy range between E and $E + dE$. Let us consider
the eigenstates $|k>$ of energy E_k of a given Hamiltonian H

$$H|k> = E_k|k> \quad . \tag{3.41}$$

Mathematically, the density of states can be defined as the following sum
of delta functions:

$$n(E) = \sum_k \delta(E - E_k) \tag{3.42}$$

An interesting form of $n(E)$ can be obtained using the resolvent or Green's
operator G, defined by [3.17,18],

$$G = \lim_{\eta \to 0^+} (E - H + i\eta)^{-1} \quad , \tag{3.43}$$

and taking the limit for infinitely small positive value of η. We can express
the trace of G (invariant against any basis change) as

$$\text{Tr}\{G\} = \lim_{\eta \to 0^+} \sum_k <k|(E - H + i\eta)^{-1}|k> \quad ,$$

---

[2] The mathematics of Green's functions and their applications to one-body
quantum problems can be found in [3.15]. A recent review of this technique
applied to point defects is given by [3.16].

that is,

$$Tr\{G\} = \sum_k \lim_{\eta \to 0^+} \frac{1}{E - E_k + i\eta} \qquad . \tag{3.44}$$

The imaginary part of this quantity is related to the total density of states $n(E)$. It is given by

$$Im\{Tr(G)\} = -\sum_k \lim_{\eta \to 0^+} \frac{\eta}{(E - E_k)^2 + \eta^2} \qquad . \tag{3.45}$$

Each term on the right-hand side of (3.45) is zero for $E \neq E_k$ and infinite for $E = E_k$. Furthermore, its integral over E is equal to π times the corresponding delta function so that

$$n(E) = -\frac{1}{\pi} Im\{Tr(G)\} \qquad . \tag{3.46}$$

We can also derive another form of $n(E)$ by simply noting that each term in the second part of (3.44) can be rewritten: $-d[\log(E - E_k + i\eta)^{-1}]/dE$ that

$$n(E) = \frac{1}{\pi} Im\left\{\frac{d}{dE} \log \prod_k \frac{1}{E - E_k + i\eta}\right\} \qquad , \tag{3.47}$$

or,

$$n(E) = \frac{1}{\pi} \frac{d}{dE} Im\left\{\log \det (G)\right\} \qquad . \tag{3.48}$$

This second form (3.48) is very useful for defect problems as we shall see later.

3.2.2 Local Densities of States and Green's Functions

Other useful quantities (for the knowledge of the charge density versus position, for instance) are the local densities of states. We shall first define this concept quite generally and then develop a particular application to tight-binding systems leading to the definition of the partial densities of states.

Any eigenfunction $\psi_k(\underline{r})$ of the Hamiltonian H is the projection of the eigenvector $|k\rangle$ onto the position vector $|\underline{r}\rangle$:

$$\psi_k(\underline{r}) = \langle \underline{r}|k\rangle \qquad . \tag{3.49}$$

We can define matrix elements of the resolvent G between two different position vectors $|r>$ and $|r'>$:

$$G(r,r') = <r|G|r'> \quad , \tag{3.50}$$

that is

$$G(r,r') = \lim_{\eta \to 0^+} \sum_k \frac{<r|k> <k|r'>}{E-E_k+i\eta} \quad . \tag{3.51}$$

The imaginary part of $G(r,r)$ is of particular interest because it is equal to

$$\text{Im}\{G(r,r)\} = - \pi \sum_k |\psi_k(r)|^2 \delta(E - E_k) \quad . \tag{3.52}$$

The right-hand term of (3.52) is proportional to what we might call the local density of states $n(r,E)$ at point r given by

$$n(r,E) = \sum_k |\psi_k(r)|^2 \delta(E - E_k) \quad , \quad \text{or} \tag{3.53}$$

$$n(r,E) = - \frac{1}{\pi} \text{Im}\{G(r,r)\} \quad . \tag{3.54}$$

One trivial property of $n(r,E)$ is that, when summed over all position vectors, i.e., when integrated over the whole space, it gives the total density of states $n(E)$, thereby justifying its label of "local density of states". A second property which is of great interest is that we can express the electron density $n(r)$ at point r in terms of this local density of states from the following relations

$$n(r) = 2 \sum_k f_k |\psi_k(r)|^2 \tag{3.55}$$

$$n(r) = 2 \int f(E)dE \sum_k |\psi_k(r)|^2 \delta(E - E_k) \tag{3.56}$$

$$n(r) = 2 \int f(E)n(r,E)dE \quad , \tag{3.57}$$

where f_k or $f(E)$ are equal to unity for occupied states, zero for empty states.

These definitions are helpful for self-consistent calculations where, as we shall see in the next chapter, we have to calculate the charge density versus the potential and so on. They show that all quantities of interest can be expressed in terms of the local Green functions $G(r,r)$.

In tight-binding calculations however, the evaluation of local Green's functions is not so essential because we work in a restricted basis set.

Instead of projecting onto position vectors it is sufficient and simpler to project onto an atomic state $|\mu>$ centered on one atom. This defines a partial density of states $n_\mu(E)$ corresponding to the μth atomic state,

$$n_\mu(E) = \sum_k |<\mu|k>|^2 \delta(E - E_k) \quad , \tag{3.58}$$

which can again be transformed in terms of a Green's function,

$$G_{\mu\mu} = <\mu|G|\mu> \quad , \tag{3.59}$$

as

$$n_\mu(E) - \frac{1}{\pi} \, \text{Im}\{<\mu|G|\mu>\} \quad . \tag{3.60}$$

Again the total density of states is the sum of all possible $n_\mu(E)$. We calculate the electron population N_μ on the μth orbital as

$$N_\mu = 2 \int f(E)n_\mu(E)dE \quad . \tag{3.61}$$

This concept of "local" or "partial" densities of states is useful and particularly suited to calculations on complicated systems using methods of moments (Sect.3.2.5).

3.2.3 Green's Function Treatment of Local Perturbations

We now apply the above definitions to the case of deep-level centers, i.e., to the problem of a perturbation strongly localized in space near the defect.

Calling H_0 the perfect crystal Hamiltonian and $H = H_0 + V$ the perturbed Hamiltonian, V being the localized perturbative potential, we have to solve the Schrödinger equation

$$(E - H_0)\psi = V\psi \quad . \tag{3.62}$$

This is a second-order differential equation having a second member. Its general solution ψ is the sum of the general solution ψ_0 of the equation without second member, i.e.,

$$(E - H_0)\psi_0 = 0 \tag{3.63}$$

and of one particular solution of the equation with second member, for instance,

$$\psi = (E - H_0)^{-1}\psi \quad . \tag{3.64}$$

When E falls within an allowed energy band of the unperturbed system, we have

$$\psi = \psi_0 + G_0 V\psi \quad . \tag{3.65}$$

In case where no solutions of H_0 exist at the energy E considered (for instance, in the forbidden gap of the semiconductor) we have then

$$\psi = G_0 V\psi \quad . \tag{3.66}$$

Any eigenstate obtained from (3.66) is a localized state whose wave function must decrease exponentially away from the defect site. If we express ψ within a given basis set (Wannier or atomic functions, or any other basis set), then we obtain a linear homogeneous system of equations for the coefficients of the expansion. This has nonvanishing solutions only if

$$\det |I - G_0 V| = 0 \quad , \tag{3.67}$$

where I is the unit matrix.

This is the equation giving the bound energy levels associated with the localized states. A first important advantage of (3.67) is that it involves the resolvent operator of the perfect crystal which can be calculated, once and for all, for any possible defect problem. The second essential simplification comes from the assumption that V is localized in the vicinity of the defect. If one uses a basis of localized functions, the matrix V takes the following form

$$\begin{bmatrix} V_{11} & 0 \\ \hline 0 & 0 \end{bmatrix} , \tag{3.68}$$

where V_{11} is a small submatrix containing the non-vanishing matrix elements of V. These elements connect only a small number of basis states located in the vicinity of the defect. From this, the matrix $I - G_0 V$ takes the following form:

$$\begin{bmatrix} I_{11} - (G_0)_{11} \, V_{11} & 0 \\ \hline -(G_0)_{21} \, V_{11} & I_{22} \end{bmatrix} , \tag{3.69}$$

the index 2 referring to the supplementary subspace of 1, I_{11} and I_{22} being
the unit matrices in each subspace.

It is now easy to show that

$$\det\left(I - G_0V\right) = \det\left[I - (G_0)_{11}V_{11}\right] \quad . \tag{3.70}$$

This means that, for localized perturbations, the bound state equation reduces
to the calculation of a fairly small determinant, allowing exact calculation
of the position of the bound state levels.

In cases where E belongs to an allowed energy band of the perfect crystal,
we have to use (3.65) for the wave function, which describes the scattering
of the Bloch states by the defect. The density of states n(E) of the perturbed
system becomes different from $n_0(E)$, the perfect-crystal density of states,
and we have to evaluate its change $\delta n(E)$. This change can readily be expressed
using (3.48) in terms of the resolvent operators G_0 for the perfect crystal
and G for the perturbed system:

$$\delta n(E) = \frac{1}{\pi} \, \text{Im} \left\{ \frac{d}{dE} \log \frac{\det(G)}{\det(G_0)} \right\} \quad . \tag{3.71}$$

Then, expressing G in terms of G_0, we write

$$G = \frac{1}{E-H_0-V+i\eta} \tag{3.72}$$

$$G = \frac{1}{E-H_0+i\eta} \left\{ 1 + \left[(E - H_0 + i\eta) - (E - H_0 - V + i\eta) \right] \frac{1}{E-H_0-V+i\eta} \right\} \quad , \tag{3.73}$$

that is

$$G = G_0 + G_0VG \quad . \tag{3.74}$$

This so-called Dyson's equation can be transformed in the following way:

$$G = (I - G_0V)^{-1}G_0 \quad , \tag{3.75}$$

which injected in (3.71), leads to

$$\delta n(E) = -\frac{1}{\pi} \frac{d}{dE} \, \text{Im} \left\{ \log \det |I - G_0V| \right\} \quad . \tag{3.76}$$

Again, the essential quantity represented by the change in density of states
$\delta n(E)$ is expressed in terms of the reduced determinant of $(I - G_0V)_{11}$ for the
same reasons as above.

A general sum rule can be derived for $\delta n(E)$ from the conservation of the
total number of states. As the number of localized states solutions of (3.67)

must be an integer N, it is clear that the integral of $\delta n(E)$ over all bands must be equal and opposite to this integer, i.e.,

$$\int \delta n(E)dE = -N \quad .$$
(3.77)

It is worth mentioning the analogy of this problem with standard scattering theory of free electrons by a localized spherically symmetric potential. We shall not develop this analogy here, but simply note that it is based on the definition of generalized phase shifts [3.19] whose sum $\delta(E)$ is given by

$$\delta(E) = - \tan^{-1} \frac{Im\{det(I-G_0 V)\}}{Re\{det(I-G_0 V)\}}$$
(3.78)

(Re denotes the real part) related to $\delta n(E)$ by

$$\delta n(E) = \frac{1}{\pi} \frac{d\delta(E)}{dE} \quad .$$
(3.79)

Thus, the total phase shift $\delta(E)$ is simply equal to π times $\delta N(E)$ the change in total number of states having an energy lower than E.

Let us finally mention that $det(I - G_0 V)$ can also vanish within an allowed energy band. The phase shift will then take a value equal to $(2n+1)\pi/2$. This is what is called a "resonance" or "antiresonance" as we shall discuss in the next section.

3.2.4 Application to the Koster-Slater Problem

We consider here the application of Green's function theory to the simplest case, i.e., the substitutional impurity in a one-band tight-binding system, the well-known Koster-Slater model [3.20]. We detail this model because it illustrates several fundamental aspects. Let us then consider a homogeneous perfect crystal with one s orbital per atom. We take the free-atom energy as the origin and, for simplicity, consider only the case of nearest neighbors interactions. We also restrict ourselves to the case of symmetrical bands, e.g., the simple cubic or the cubic centered lattice, but not the face centered cubic one. However, all these simplifying assumptions are not essential.

As we have shown before, the density of states per atom of the unperturbed crystal has the schematic form of Fig.3.16a. We label it $n_0(E)$ because it is equal to the local density of states $n_\mu(E)$ on each atom and particularly on the atom located at the origin. This density of states can be calculated by standard numerical methods. Once this is done, we know the imaginary part of each Green's function $<i|G_0|i>$ because

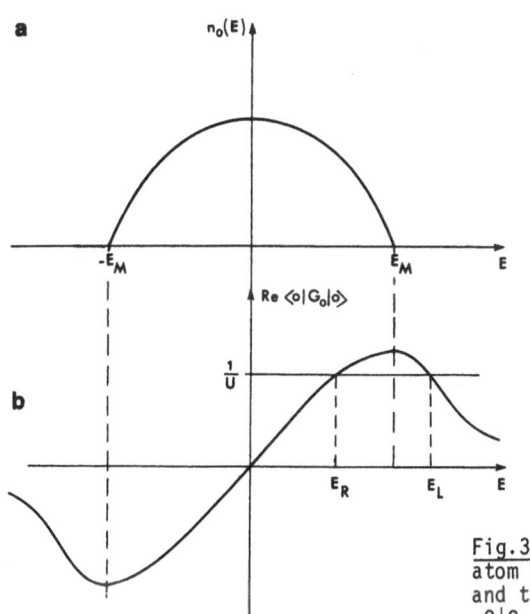

Fig.3.16a,b. Density of states per atom in the unperturbed crystal (a) and the corresponding real part of $<0|G_0|0>$ (b) versus energy. See text for the notations

$$\text{Im}\{<i|G_0|i>\} = \text{Im}\{<0|G_0|0>\} = -\pi n_0(E) \tag{3.80}$$

From (3.80) it is not difficult to calculate the real part $\text{Re}\{<0|G_0|0>\}$ noticing that

$$\text{Re}\{<0|G_0|0>\} = \text{Re}\left\{\lim_{\eta\to 0^+} \sum_k \frac{|<0|k>|^2}{E-E_k+i\eta}\right\} \tag{3.81}$$

$$\text{Re}\{<0|G_0|0>\} = \lim_{\eta\to 0^+} \sum_k \frac{(E-E_k)|<0|k>|^2}{(E-E_k)^2+\eta^2} \quad . \tag{3.82}$$

This quantity is equal to $(E - E_k)^{-1}$ for $E \neq E_k$ and zero for $E = E_k$. It can also be noted that

$$\text{Re}\{<0|G_0|0>\} = \mathscr{P} \sum_k \frac{|<0|k>|^2}{E-E_k} \quad , \tag{3.83}$$

where $\mathscr{P}$ denotes the principal part, which is related to the local density of states

$$n_0(E) = \sum_k |<0|k>|^2 \delta(E - E_k) \quad , \tag{3.84}$$

through the Hilbert transform

$$Re\{<0|G_0|0>\} = \mathscr{P} \int \frac{n_0(E')dE'}{E-E'} \quad . \tag{3.85}$$

This integral relation allows the calculation of $Re\{<0|G_0|0>\}$, whose schematic behavior is given in Fig.3.16b. For a symmetric density $n_0(E)$, $Re\{<0|G_0|0>\}$ is antisymmetric and thus equal to zero at $E = 0$. Another important property is that this real part tends to zero as $1/E$, when E tends to infinity [as can be seen from (3.85)]. In the following, for simplicity, we shall denote as $R_0(E)$ and $I_0(E)$ the real and imaginary parts of $<0|G_0|0>$.

Let us now put a substitutional impurity at site 0, such that only the atomic energy at that site is shifted by a quantity U, the other interatomic interactions remaining unchanged. As shown by (3.67,76) the central quantity is given by $\det|I - G_0V|$, which in this case takes the following simple form

$$\det|I - G_0V| = 1 - <0|G_0|0> U \quad . \tag{3.86}$$

The equation for localized states which must lie outside the band, i.e., for $|E|>|E_M|$ (see Fig.3.16a) thus becomes

$$R_0(E) = 1/U \quad , \tag{3.87}$$

because in this range of energy the density of states $n_0(E)$ and thus $I_0(E)$ vanish. Equation (3.87) has roots only if $|U|$ exceeds a certain maximum value. For repulsive potentials $U > 0$, a bound state is raised above the band, while for attractive potentials $U < 0$, it is lowered under the band in a symmetrical way. Figure 3.16b shows the corresponding position E_L of the localized state in the repulsive case.

Another very important case corresponds to $Re\{\det(I - G_0V)\}$ vanishing within the band. We have seen, from (3.78), that the denominator occuring in the definition of the phase shift becomes equal to zero, so that $\delta(E)$ becomes equal to $(2n + 1)\pi/2$ and the $\tan^{-1}$ can change its determination. In any case it is of interest to analyze the behavior of $\delta(E)$ in this energy range. For this we apply (3.78) in the vicinity of $E = E_R$ (Fig.3.16b) where this singularity occurs. In this case $\delta(E)$ is given by

$$\delta(E) = \tan^{-1} \frac{I_0(E)U}{1-R_0(E)U} \quad , \tag{3.88}$$

Around the point $E = E_R$ we can approximate $I_0(E)$ by $I_0(E_R)$ and replace

$1 - R_0(E)U$ by $-R_0'(E_R)U(E - E_R)$ where $R_0'(E)$ is the derivative of $R_0(E)$ with respect to the energy. We thus find that

$$\delta(E) \simeq - \tan^{-1} \frac{I_0(E_R)}{R_0'(E_R)(E-E_R)} . \tag{3.89}$$

This allows calculation of $\delta n(E)$ for $E \simeq E_R$. We find

$$\delta n(E) = \frac{\Gamma}{2\pi} \frac{1}{(E-E_R)^2 + \frac{\Gamma^2}{4}} , \tag{3.90}$$

with

$$\Gamma = \frac{2I_0(E_R)}{R_0'(E_R)} . \tag{3.91}$$

Equation (3.90) represents a Lorenzian curve. Thus, $\delta n(E)$ will develop a sharp peak near E_R if the width Γ is very small, i.e., in regions where $I_0(E_R)$ is weak compared to $R_0'(E_R)$ or in regions of weak densities of states. This is a resonance when $\Gamma > 0$ and an antiresonance when $\Gamma < 0$, which is obviously the case here.

In order to see this more clearly, we have plotted in Fig.3.17 the behavior of $\delta(E)$ and $\delta n(E)$ for the repulsive potential of Fig.3.16, which leads both to a localized state and an antiresonance. The behavior of $\delta(E)$ is governed by (3.88) within $n\pi$. The value of n is imposed by the continuity of $\delta(E)$. Let us explain this in detail: first for $E < -E_M$, $\delta(E)$ is zero; then

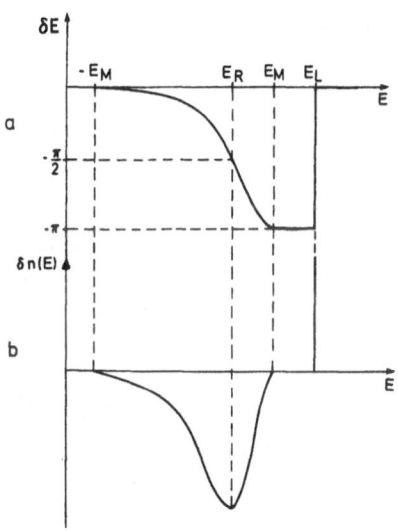

Fig.3.17.a,b. Total phase shift versus energy (a) and change in the density of states due to the introduction of the impurity (b)

for $E > -E_M$, $\delta(E)$ becomes negative until E reaches E_R where it becomes equal to $-\pi/2$; for $E = E_R$, $\tan^{-1}$ is discontinuous so that for $E > E_R$ we must now take the determination with $n = -1$ to get continuity of $\delta(E)$; thus between E_R and E_M, $\delta(E)$ still decreases down to $-\pi$ at the band limit where the $\tan^{-1}$ is zero. That this behavior is correct can be judged from $\delta N(E)$, the change in total number of states of energy smaller than E, equal to $\delta(E)/\pi$. This quantity must be equal to -1 at $E = E_M$, corresponding to the localized state which is issued from that band. Finally, taking $(1/\pi)d\delta(E)/dE$, we obtain $\delta n(E)$ pictured in the second part of the figure.

3.2.5 Green's Function and the Method of Moments

We can now introduce the methods based on the expansion of the resolvent operator G in series, i.e., on the evaluation of the moments of the density of states. We write

$$G = \frac{1}{E} \left(I - \frac{H}{E} \right)^{-1} \quad , \tag{3.92}$$

which leads to

$$G = \sum_n \frac{H^n}{E^{n+1}} \quad . \tag{3.93}$$

The energy E contains a $i\eta$ term, so we implicitly take the limit of E for $\eta \to 0^+$. We assume that this power series expansion converges, i.e., that E lies sufficiently outside the spectrum of eigenvalues of H so that the convergence conditions are fulfilled. From (3.92) any partial density of states can be written:

$$n_\alpha(E) = -\frac{1}{\pi} \, \text{Im} \left\{ \sum_{n=0}^{\infty} \frac{\langle \alpha | H^n | \alpha \rangle}{E^{n+1}} \right\} \quad , \tag{3.94}$$

that is

$$n_\alpha(E) = -\frac{1}{\pi} \, \text{Im} \left\{ \sum_{n=0}^{\infty} \frac{\mu_{n\alpha}}{E^{n+1}} \right\} \quad . \tag{3.95}$$

The $\mu_{n\alpha}$ defined as

$$\mu_{n\alpha} = \langle \alpha | H^n | \alpha \rangle \quad , \tag{3.96}$$

are the moments of the partial density of states $n_\alpha(E)$ since

$$\mu_{n\alpha} = \sum_k |\langle \alpha | k \rangle|^2 E_k^n \quad , \tag{3.97}$$

or

$$\mu_{n\alpha} = \int dE \sum_k |<\alpha|k>|^2 \delta(E - E_k)E_k^n \quad , \tag{3.98}$$

and finally

$$\mu_{n\alpha} = \int n_\alpha(E)E^n dE \quad . \tag{3.99}$$

These moments $\mu_{n\alpha}$ can be expressed under the form

$$\mu_{n\alpha} = \sum_{\alpha_1\alpha_2\cdots\alpha_{n-1}} <\alpha|H|\alpha_1> <\alpha_1|H|\alpha_2> \cdots <\alpha_{n-1}|H|\alpha> \quad . \tag{3.100}$$

The calculation of the moments is equivalent to the determination of the number of possible different circuits of n walks authorized by the Hamiltonian matrix elements, starting and ending on the given orbital $|\alpha>$ [3.21]. In simple cases this can be evaluated from geometrical considerations. In "realistic" situations the $\mu_{n\alpha}$ have to be determined numerically. The advantage is that it is a direct space calculation which can be done for a perturbed system (it has been, for instance, applied to disordered systems).

Once the moments are known it is necessary to reconstruct the density of states $n_\alpha(E)$. For this, a powerful method consists in expressing the Green's function $<\alpha|G|\alpha>$ as a continued fraction expansion, i.e., under the form

$$<\alpha|G|\alpha> = \cfrac{1}{E+a_1 - \cfrac{b_1}{E+a_2 - \cfrac{b_2}{E+a_3\ldots}}} \tag{3.101}$$

The coefficients a_i, b_i of this continued fraction are related to the moments $\mu_{n\alpha}$ by well-known expressions [3.22,23]. In practice only a limited number of these coefficients can be calculated, up to a_p and b_p, for instance. One method is to stop the continued fraction at this level p. This leads to an expression ,

$$<\alpha|G|\alpha> = \sum_{i=1}^p \frac{\omega_i}{E-E_i} \quad , \tag{3.102}$$

which is equivalent to say that the density of states $n_\alpha(E)$ is the sum of p delta functions. From such a density of states, smoothing procedures can be used to lead to a continuous curve, approximating the true density of states. Good approximations require usually p to be about 10 to 15.

Another method of terminating the continuous fraction is to assume that at level p, the coefficients a_i and b_i have reached a limiting value $a = a_p$ and $b = b_p$. In this case we can write [3.23]

$$<\alpha|G|\alpha> = \cfrac{1}{E-a_1 - \cfrac{b_1}{E.... \quad \cfrac{}{E-a_p-b_p G}}} \quad , \tag{3.103}$$

with

$$G = \frac{1}{E-a-bG} \quad , \tag{3.104}$$

or

$$G = \frac{E-a-\sqrt{(E-a)^2-4b}}{2b} \quad . \tag{3.105}$$

Such a termination is only correct in particular cases. From (3.103) G can only become imaginary in one energy interval so that we obtain only one band. However, if p is sufficiently high, we can obtain gaps as regions of very low densities of states.

The a_i and b_i can be determined directly by putting H under a tridiagonal form, i.e., transforming the Hamiltonian matrix into a semiinfinite linear chain for which the continuous fraction expansion follows immediately. Again we want to calculate $<\alpha|G|\alpha>$ so that we start from $|\alpha>$. We can write

$$H|\alpha> = \sum_{i\neq\alpha} H_{\alpha i}|i> + \mu_{1\alpha}|\alpha> \quad , \tag{3.106}$$

where $\mu_{1\alpha}$ is the first moment of $n_\alpha(E)$, equal to $H_{\alpha\alpha}$ from (3.96). We now define the normalized vector $|\alpha_1>$ by

$$|\alpha_1> = \frac{1}{\sqrt{\mu'_{2\alpha}}} \sum_{i\neq\alpha} H_{\alpha i}|i> \quad , \tag{3.107}$$

with

$$\mu'_{2\alpha} = \sum_{i\neq\alpha} |H_{\alpha i}|^2 \quad . \tag{3.108}$$

Here $\mu'_{2\alpha}$ is the contribution of interatomic interactions to the second moment of $n_\alpha(E)$. We can now take as basis states the two states $|\alpha>$ and $|\alpha_1>$ and then all orthonormal combinations of the remaining states. The corresponding Hamiltonian matrix will contain the following elements

$$H_{\alpha\alpha} = \mu_{1\alpha}, \quad H_{\alpha\alpha 1} = H_{\alpha 1\alpha} = \sqrt{\mu'_{2\alpha}} \quad , \quad H_{\alpha i} = H_{i\alpha} = 0 \quad , \quad \text{for } i \neq \alpha \text{ and } \alpha_1 \quad . \tag{3.109}$$

We can repeat the same procedure replacing $|\alpha\rangle$ by $|\alpha_1\rangle$ and $|\alpha_1\rangle$ by $|\alpha_2\rangle$ such that

$$|\alpha_2\rangle = \frac{1}{\sqrt{\mu'_{2\alpha_1}}} \sum_{i \neq \alpha, \alpha_1} H_{\alpha_1 i} |i\rangle$$

$$\mu'_{2\alpha_1} = \sum_{i \neq \alpha, \alpha_1} |H_{\alpha_1, i}|^2 \quad . \tag{3.110}$$

This can be repeated indefinitely. It is clear that the Hamiltonian matrix will become equivalent to the linear chain of Fig.3.18, with

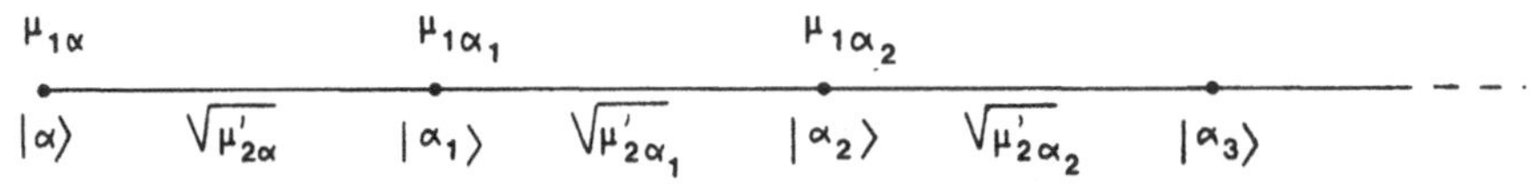

Fig.3.18. Linear chain equivalent to the continued fraction expansion

$$|\alpha_i\rangle = \frac{1}{\sqrt{\mu'_{2\alpha_i}}} \sum_{\substack{k \neq \alpha \\ \alpha_1 \\ \vdots \\ \alpha_{i-1}}} H_{\alpha_i, k}|k\rangle \; , \quad \mu_{1\alpha_i} = \langle \alpha_i|H|\alpha_i\rangle \; , \quad \mu'_{2\alpha_i} = \sum_{\substack{k \neq \alpha \\ \alpha_1 \\ \vdots \\ \alpha_i}} |H_{\alpha_i, k}|^2$$

$$\tag{3.111}$$

For a semiinfinite linear chain, the continued fraction expansion can be obtained immediately in terms of our $\mu_{1\alpha_i}$ and $\mu'_{2\alpha_i}$, as we shall see in the next section.

3.2.6 Green's Function for a Semiinfinite Chain: Application to the Vacancy

Let us calculate the Green's function $\langle\alpha|G|\alpha\rangle$ corresponding to the semiinfinite chain of Fig.3.18. For this, we define a new Hamiltonian H_1 and its resolvent G_1 corresponding to the same chain where the interaction between $|\alpha\rangle$ and $|\alpha_1\rangle$ has been suppressed. Using Dyson's equation (3.74) we can write

$$\langle\alpha|G|\alpha\rangle = \langle\alpha|G_1|\alpha\rangle + \langle\alpha|G_1|\alpha\rangle \sqrt{\mu'_{2\alpha}} \langle\alpha_1|G|\alpha\rangle \tag{3.112}$$

and

$$\langle\alpha_1|G|\alpha\rangle = \langle\alpha_1|G_1|\alpha_1\rangle \sqrt{\mu'_{2\alpha}} \langle\alpha|G|\alpha\rangle \quad . \tag{3.113}$$

The quantity $\langle\alpha|G_1|\alpha\rangle$ is the Green's function corresponding to the isolated orbital $|\alpha\rangle$ with atomic energy $\mu_{1\alpha}$. It is thus equal to $(E-\mu_{1\alpha})^{-1}$ and we can write

$$\langle\alpha|G|\alpha\rangle = \frac{1}{E-\mu_{1\alpha}-\mu'_{2\alpha}\langle\alpha_1|G_1|\alpha_1\rangle} \quad . \tag{3.114}$$

Here $\langle\alpha_1|G_1|\alpha_1\rangle$ corresponds to a new semiinfinite chain beginning on α_1. We thus repeat the same procedure, defining H_2 and G_2 such that $\mu'_{2\alpha_1}$ is suppressed. At level i, we have

$$\langle\alpha_i|G_i|\alpha_i\rangle = \left(E - \mu_{1\alpha_i} - \mu'_{2\alpha_i}\langle\alpha_{i+1}|G_{i+1}|\alpha_{i+1}\rangle\right)^{-1} \quad . \tag{3.115}$$

Using (3.114,115) we write the general expression,

$$\langle\alpha|G|\alpha\rangle = \cfrac{1}{E-\mu_{1\alpha} - \cfrac{\mu'_{2\alpha}}{E-\mu_{1\alpha_1} - \cfrac{\mu'_{2\alpha_1}}{\cfrac{----}{E - \mu_{1\alpha_i} - \cfrac{\mu'_{2\alpha_i}}{E---}}}}} \quad , \tag{3.116}$$

which identifies a_i with $\mu_{1\alpha_i}$ and b_i with $\mu'_{2\alpha_i}$.

An interesting application of this is provided by the vacancy in the covalent linear chain. This leads to the semiinfinite problem of Fig.3.19a.

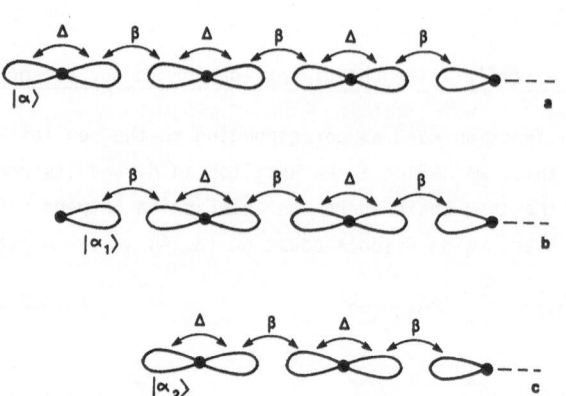

Fig.3.19. The semiinfinite linear chain (a). The first step (b) leads to (3.117) and the second one (c) to (3.118)

There, the $\mu_{1\alpha_i}$ and $\mu'_{2\alpha_i}$ are immediately defined. All $\mu_{1\alpha_i}$ are zero and $\sqrt{\mu'_{2\alpha_i}}$ is equal either to $|\Delta|$ or to $|\beta|$; $|\alpha\rangle$ is the dangling orbital. We can thus write

$$\langle\alpha|G|\alpha\rangle = \frac{1}{E-\Delta^2\langle\alpha_1|G_1|\alpha_1\rangle} \tag{3.117}$$

and

$$\langle\alpha|G|\alpha\rangle = \frac{1}{E - \dfrac{\Delta^2}{E-\beta^2\langle\alpha_2|G_2|\alpha_2\rangle}} \quad . \tag{3.118}$$

The expression (3.117) corresponds to Fig.3.19b and (3.118) to Fig.3.19c. Clearly, we have

$$\langle\alpha_2|G_2|\alpha_2\rangle = \langle\alpha|G|\alpha\rangle \quad , \tag{3.119}$$

so that

$$\langle\alpha|G|\alpha\rangle = \frac{E-\beta^2\langle\alpha|G|\alpha\rangle}{E(E-\beta^2\langle\alpha|G|\alpha\rangle)-\Delta^2} \quad . \tag{3.120}$$

For $\langle\alpha|G|\alpha\rangle$ this leads to a quadratic equation whose solutions are

$$\langle\alpha|G|\alpha\rangle = \frac{1}{2\beta^2 E}\left[E^2 + \beta^2 - \Delta^2 \pm \sqrt{(E^2 + \beta^2 - \Delta^2)^2 - 4E^2\beta^2}\right] \quad . \tag{3.121}$$

The sign has to be chosen so that in each energy region, $\langle\alpha|G|\alpha\rangle$ has the correct analytic behavior. We find that $\langle\alpha|G|\alpha\rangle$ has an imaginary part for $\beta+\Delta < E < \beta-\Delta$ and $-\beta+\Delta < E < -\beta-\Delta$ (if $\beta<0$, $\Delta<0$, and $|\beta|>|\Delta|$ as is the case in covalent systems). We also find that $\langle\alpha|G|\alpha\rangle$ has one simple pole at $E = 0$, which is just at the middle of the forbidden gap, and the $+$ sign in (3.121) is appropriate. The weight of this delta function is the residue of $\langle\alpha|G|\alpha\rangle$ at this simple pole, simply equal to $1-\Delta^2/\beta^2$. The local density of states $n_\alpha(E)$ on the dangling orbital therefore has the shape of Fig.3.20. This shows that there is a localized state at $E = 0$, the contribution of the dangling orbital to the bound state wave function being $1-\Delta^2/\beta^2$, and the remaining contribution Δ^2/β^2 being equally shared between the two bands.

We have chosen here to calculate the Green's function on the dangling orbital, because it illustrates a powerful method which has been used in disordered systems. We can also treat the problems using the standard methods discussed in Sect.3.2.3.

108

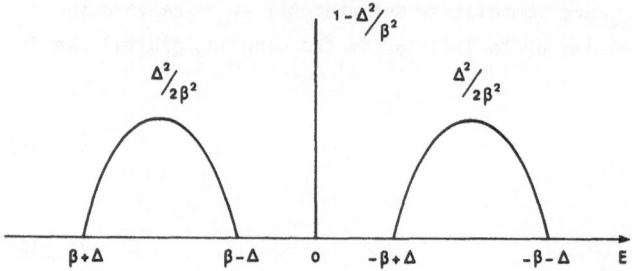

Fig.3.20. Local density of states on the dangling orbitals of a semiinfinite linear chain. See text for the notations

Another advantage of the procedure we have just used is that it applies, without change, to the problem of the vacany in a Bethe lattice of fourfold coordination, using the same Hamiltonian defined by our two parameters β, Δ. In such a lattice each dangling orbital is connected to its own lattice and can thus be treated in an independent way. The problem can be cast into the semiinfinite chain form of Fig.3.21. Again after two steps, we recover the original $\langle\alpha|G|\alpha\rangle$

$$\langle\alpha|G|\alpha\rangle = \cfrac{1}{E - \cfrac{3\Delta^2}{E-2\Delta-\beta^2\langle\alpha|G|\alpha\rangle}} \quad , \tag{3.122}$$

which gives,

$$\langle\alpha|G|\alpha\rangle = \frac{1}{2\beta^2 E}\left\{E(E-2\Delta)+\beta^2-3\Delta^2 \pm \left[\left(E(E-2\Delta)+\beta^2-3\Delta^2\right)^2 - 4\beta^2 E(E-2\Delta)\right]^{\frac{1}{2}}\right\} \quad . \tag{3.123}$$

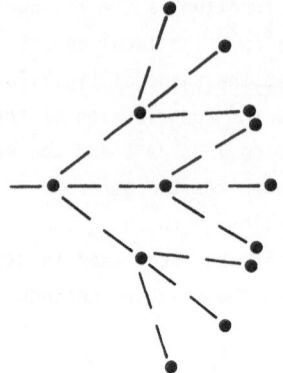

Fig.3.21. Bethe lattice for fourfold coordinated atoms

This expression is completely similar to (3.121), replacing E^2 by $E(E - 2\Delta)$, i.e., by $(E - \Delta)^2 - \Delta^2$, and Δ by $\sqrt{3}\Delta$. The density of states now extends between $\Delta - \sqrt{\Delta^2 + (\beta + \sqrt{3}\Delta)^2}$ and $\Delta - \sqrt{\Delta^2 + (\beta - \sqrt{3}\Delta)^2}$, and between $\Delta + \sqrt{\Delta^2 + (\beta - \sqrt{3}\Delta)^2}$ and $\Delta + \sqrt{\Delta^2 + (\beta + \sqrt{3}\Delta)^2}$. For $|\beta| > \sqrt{3}|\Delta|$ there is a bound state at $E = 0$, whose weight $1 - (3\Delta^2/\beta^2)$ gives the measure of the delocalization of the bound-state wave function. The final localized level is fourfold degenerate corresponding to the four dangling orbitals.

3.2.7 Refined Tight-Binding Green's Functions Treatments

We describe here results concerning the single vacancy. The most complete treatment for this defect uses the formalism of Sect.3.2.3 in which the central quantity is the generalized phase shift defined by (3.78). To apply this simple formalism with the minimal amount of computational work, the perturbation is defined in the following way. A symmetrical basis is taken on the central atom, consisting of its s orbital and its three p orbitals. In an environment having the symmetry T_d, χ_{0s} forms a basis for the A_1 representation, χ_{0p_x}, χ_{0p_y}, and χ_{0p_z} a basis for T_2 (see Sect.1.4.3). These four orbitals will not be mixed by any symmetrical operator such as H_0 or G_0, the Hamiltonian and resolvent of the perfect crystal. Let us now define a fictitious perturbation U having only the four intraatomic matrix elements

$$\langle \chi_{0i} | H | \chi_{0j} \rangle = U\delta_{ij} \quad , \tag{3.124}$$

corresponding to some sort of substitutional impurity placed at the origin. In this case, it is easy to calculate $\det (I - G_0 V)$

$$\det \left(I - G_0 V \right) = \left(1 - U \langle \chi_{0s} | G_0 | \chi_{0s} \rangle \right) \left(1 - U \langle \chi_{0p_x} | G_0 | \chi_{0p_x} \rangle \right)^3 \tag{3.125}$$

because the contribution of χ_{0p_y} and χ_{0p_z} is the same as for χ_{0p_x}. From this, the bound-state energies are given by

$$\langle \chi_{0s} | G_0 | \chi_{0s} \rangle = 1/U \quad (A_1)$$

$$\langle \chi_{0p_x} | G_0 | \chi_{0p_x} \rangle = 1/U \quad (T_2) \quad . \tag{3.126}$$

We have now to choose the value of U, if any, which can be used to simulate the vacancy. This can be obtained by taking the limit $U \to \infty$, in which case the atomic levels of the central site atom are raised infinitely high in energy so that they can no longer interact with the states of the remaining crystal. The possible bound states for the vacancy are then

$$<x_{0s}|G_0|x_{0s}> = 0$$

$$<x_{0p_x}|G_0|x_{0p_x}> = 0 \quad . \tag{3.127}$$

The same definition applies to the calculation of $\delta N(E)$, the change in the total number of states of energy less than E, equal to $\delta(E)/\pi$ according to (3.79). We obtain

$$\delta N(E) = -\frac{1}{\pi}\left(\tan^{-1}\frac{<x_{0s}|I_0|x_{0s}>}{<x_{0s}|R_0|x_{0s}>} + 3\tan^{-1}\frac{<x_{0p_x}|I_0|x_{0p_x}>}{<x_{0p_x}|R_0|x_{0p_x}>}\right) , \tag{3.128}$$

I_0 and R_0 standing for the imaginary and real parts of G_0.

All results of interest for this tight-binding treatment of the vacancy are obtained from the knowledge of two intraatomic matrix elements of G. These elements have been calculated using many models; the first one [3.24] includes only the two interactions β and Δ. In this case a fourfold degenerate localized state is found at the energy E = 0, exactly as in Sect.3.2.6 for the same model in a Bethe lattice. This is not surprising in view of the fact that this model does not incorporate the second neighbors' interaction γ between dangling orbitals, which is responsible for the splitting of the localized state. However, it is worth reproducing $\delta N(E)$ derived from this model; it is given in Fig.3.22, where the s and p contributions have been drawn separately. This figure shows that one s state and three p states are lost in the valence band and are just compensated by the four bound states located in the gap. The same situation occurs in the conduction band where four states, corresponding to the states of the missing atom which here have been repelled to infinity, are lost.

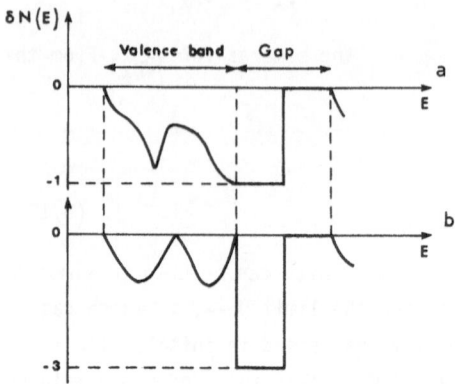

Fig.3.22a,b. Change in the density of states due to the introduction of a vacancy in a model including the interactions β and Δ: (a) s part; (b) p part

It is important to note that since the valence band now only contains 2N - 4 states (if N is the number of atoms in the perfect crystal), it can only accomodate 4N - 8 electrons. For the neutral vacancy case (corresponding to the removal of a neutral atom), the system contains 4N - 4 electrons. We thus have to fill the valence band and put four electrons on the localized level. This is exactly what we have found in the molecular model.

A more refined tight-binding calculation has been made in [3.5,25]. The authors have evaluated $<x_{0s}|G_0|x_{0s}>$ and $<x_{0p_x}|G_0|x_{0p_x}>$ by the continued fraction technique and, as we have discussed above, obtained the fundamental energy gap as a region of extremely low density of states. They also include interactions between second nearest neighors and thus obtain a splitting of the A_1 and T_2 localized states. They find that for silicon, in a two-center approximation, the A_1 state falls within the valence band at -0.88 eV from its top and the T_2 state falls within the gap at +0.12 eV from the top of the valence band. One problem is that the results depend on the parametrization scheme (the above values change to -1.10 and 0.42 when using a three-center approximation). For comparison with the simplified version of Fig.3.22, we give in Fig.3.23 the corresponding $\delta N(E)$. They exhibit essentially the same behavior apart from some shifts in energy. The arguments for the electron population of levels have only to be changed slightly. The A_1 state, which is lower, is always filled with two electrons even if it falls within the valence band. Thus only two electrons remain available for the T_2 level which must then fall within the gap. It is also interesting to note that the same model, applied to diamond, leads to A_1 and T_2 at 0 and 1.8 eV from the top of the valence band.

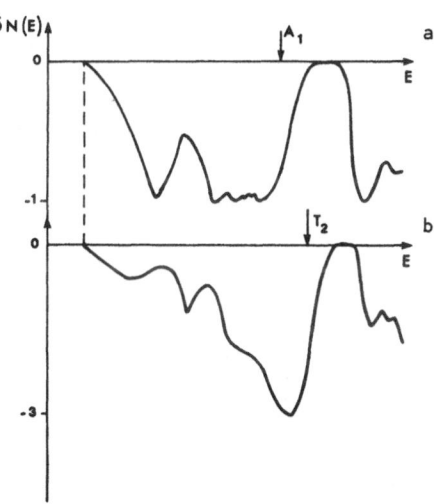

Fig.3.23a,b. Change in the density of states introduced by a vacancy according to [3.25]. (a) s part; (b) p part

Essentially similar results have been found for silicon in [3.26]. Here the authors computed directly the Green's functions by direct summation over the Brillouin zone of the perfect crystal, thereby obtaining a well-defined value for the gap. They find a T_2 level at +0.27 eV.

We shall close this section by a discussion of results [3.27] for the tetrahedral interstitial, which we had discussed qualitatively using the molecular model. For silicon, there are no levels within the forbidden gap, but a A_1 localized level under the valence band. These results exactly correspond to the discussion we have given in Sect.3.1.6. In diamond, the situation is slightly different: A_1 and T_2 states are found within the gap (at + 0.32 and 1.32 eV from the valence band), a A_1 state under the valence band and a T_2 localized above the conduction band. Again these results correspond to our previous discussion, but now, the $A_1 - T_2$ splitting is smaller than the fundamental gap so that these states are not found as resonances in the valence and conduction bands.

Finally, we want to stress the important point that the tight-binding calculations, which have been described in this chapter, are semiempirical and cannot give by themselves exact a priori results for the defect states. We shall see in the next chapter how these limitations have been overcome recently and how Green's functions techniques can be applied to quantitative self-consistent calculations.

4. Many-Electron Effects and Sophisticated Theories of Deep Levels

Any quantitative theory of deep levels in semiconductors must incorporate the electron-electron interactions. In the simplified tight-binding treatments we have described in Chap.3, these effects have been ignored, i.e., we have assumed an independent electron approximation.

The exact Hamiltonian of the electronic system is a complex differential operator that cannot be written as a sum of independent one-electron Hamiltonians in view of the electron-electron coupling. There is, at the moment, no method of dealing exactly with this problem, so we must use approximations. All the approximations are based on the variational principle, starting from a trial wave function written in terms of one-electron wave functions which are the variational parameters. Minimization of the total energy with respect to these one-electron wave functions leads to a set of one-electron-like equations allowing the numerical determination of these wave functions. Such a procedure is clearly valid for the ground state and sometimes for excited states (when orthogonality with the exact ground state is automatically insured, i.e., in most cases, by symmetry). However, its application to the determination of excitation energies is much less trivial.

The simplest of these variational procedures is the well-known Hartree approximation. The trial wave function is written as a simple product of one-electron wave functions. The effective one-electron Hamiltonians obtained in this way contain the kinetic energy, plus the potential energy of interaction with the nuclei, and finally, the interaction of the electron in question with the charge density of all other electrons, leading to a pure problem of electrostatics. One difficulty is that the problem must be solved in a self-consistent manner because each effective one-electron Schrödinger equation contains the unknown wave functions in the charge density term. Usually this is solved using an iterative procedure.

The step beyond this is the Hartree-Fock approximation. Here, the total wave function is expressed as a Slater determinant (the most general combi-

nation of simple products satisfying the antisymmetry of the wave function)
of one-electron spin orbitals. The variational procedure leads to effective
one-electron equations in which the Hamiltonian is simply a Hartree Hamiltonian reduced by an "exchange term". Unfortunately, the Hartree-Fock solutions are often worse than those of the Hartree type for many physical properties, but not for the ground-state energy. Another important point is that
the corresponding one-electron orbitals are now orthogonal.

Corrections to the exact Hartree-Fock solution are the "correlation"
effects. They are often treated by using the configuration interaction method. For this, Slater determinants that differ from the ground-state determinant are built by allowing one or several electrons to be in a higher energy one-electron spin orbital. The exact Hamiltonian matrix is then diagonalized in the basis formed by all these determinants. In this way it is possible,
in principle, to obtain exact solutions to the many-electron problem. In practice, there are evident computational limitations to such a method.

At this stage we can also mention that there are still better one-electron
approximations, which generate the ground-state wave function from a set of
one-electron spin orbitals by more elaborate techniques [4.1] and thus automatically include some of the correlation effects. Of a different nature are
the approximations derived from the density functional formalism of HOHENBERG
and KOHN [4.2]. There it was shown that the ground-state energy can be written
as a universal functional of the electron density, which should be minimum
for the correct density. Application of this variational principle demonstrated that the problem is strictly equivalent to a set of independent one-electron equations with an effective exchange-correlation potential. The problem
is that the general form of this potential is unknown. Usually, it is replaced
by a simplified local approximation derived from the free-electron gas limit.
This leads to the "local-density approximation". To this class of approximations belongs the X_α method for which a special local from of the exchange
potential is used [4.3].

The aim of this chapter is to describe the attempts which have been made
to deal with these many-electron effects, especially in the case of the single
vacancy in diamond or silicon. In the first part, we shall describe self-
consistent one-electron calculations, starting with simplified treatments
from which semiquantitative conclusions can be drawn and ending with the
most sophisticated recent calculations applying Green's functions techniques
to a "local density" formalism. We shall also present a Thomas-Fermi calculation, which allows reproduction of these results with surprising accuracy,
shedding some light on their general nature. The second part of the chapter

will be devoted to a discussion of correlation effects in terms of "config-uration interaction". These effects will be introduced for a simplified de-scription of the hydrogen molecule. The same model will then be applied to the study of the different charge states of the vacancy within the "defect-molecule approximation". From this we shall try to draw general conclusions about the applicability of one-electron techniques to the problem of "deep levels".

4.1 One-Electron Self-Consistent Calculations

In a Hartree description, any point defect will create a disturbance such that the electron density will be modified from its perfect crystal reparti-tion. Regions with accumulation or depletion of charge induce an electro-static potential which must be included in the treatment in a self-consistent manner. The same conclusions hold for more refined techniques such as the Hartree-Fock or the "local density" formalism. To illustrate this effect, we use an extension of the tight-binding theory, analogous in some respects to the CNDO theory (see Sect. 4.1.3) of quantum chemistry.

4.1.1 Charge-Dependent Tight-Binding Treatment

As we have seen in Chap.3, the tight-binding eigenfunction ψ_k of energy E_k is written as a linear combination of atomic orbitals $\chi_{i\alpha}$ (i is the atomic index and α the orbital index), assumed to form an orthonormal basis set

$$\psi_k = \sum_{i,\alpha} a_{i\alpha}^k \chi_{i\alpha} \quad . \tag{4.1}$$

Once the eigenvalues and eigenvectors of the Hamiltonian matrix have been determined, we fill the levels with electrons and finally calculate the num-ber of electrons M_i on the ith atom as

$$M_i = \sum_k f(k) \sum_\alpha |a_{i\alpha}^k|^2 \quad , \tag{4.2}$$

where $f(k)$ is equal to one for occupied states; the sum over k includes the spin.

For nonuniform situations such as those that occur near a defect, the atomic populations M_i can be different from the neutral-free-atom values Z_i, i.e., each atom will bear a net charge. It is natural to think that the Hamil-tonian matrix elements will be functions of these charges. Usually such a dependence is neglected in "semiempirical" or "realistic" calculations, which are not self-consistent and thus cannot deal with the charge-state dependence of a defect. The self-consistency can be incorporated in a simple manner. It is coherent with the principles of the tight-binding theory to assume that only the diagonal terms are charge dependent. These terms can then be written as

$$H_{i\alpha,i\alpha} = H^0_{i\alpha,i\alpha} + \sum_j (M_j - Z_j)\gamma_{ij} \quad , \tag{4.3}$$

where $H^0_{i\alpha,i\alpha}$ is the matrix element in the absence of net charges $M_j - Z_j$. The second term corresponds to the potential energy of interaction between an electron on atom i and this excess charge. It is taken to be independent of the orbital index α for reasons of invariance in a change of basis set. The γ_{ij} are intra or interatomic Coulomb terms which can either be calculated or determined semiempirically [4.4]. For instance, the interatomic γ_{ij} tend to $1/R_{ij}$, the inverse of the interatomic distance for distant atoms.

Equations (4.1-3) require a self-consistent determination of the solutions. This has been done in perfect crystals such as III-V and II-VI compounds to relate the atomic charges with ionicity [4.5]. Applications have been made to the surfaces of these compounds and particularly the polar <111> surfaces for which electrostatic effects are essential. Finally, there has been one calculation of the properties of point defects along these lines [4.6] using the molecular model for simplicity. The most interesting conclusion of this last calculation is that charge-state effects are such that the distance in energy between the positive charge state V^+ of the vacancy in silicon and its negative one V^- is of the order of the energy gap. This is in rough agreement with the experimental data in silicon, where the energy levels of V^0, V^+, and V^- are found within the forbidden gap.

A qualitative discussion of charge effects is best illustrated for the case of a substitutional isoelectronic impurity. As we have seen in Chap.3, the main perturbation in that case is the change in sp^3 average energy which we had called U. Clearly, this will polarize the neighboring bonds. For instance, if $U < 0$ (attractive potential), there will be accumulation of electrons on the impurity. This, in turn, will induce a potential having a repulsive character on the impurity site, the final situation being self-consistent. In practice, the final value for the perturbative potential is found to be much smaller than the original or "bare" value. The reason is that the potential is screened by the electron cloud, the electron reaction tending to reduce the original perturbation. Here the main effect is to reduce U to a final value U_{eff}, much lower in magnitude.

4.1.2 The Model of HALDANE and ANDERSON

A similar model has been used by HALDANE and ANDERSON to discuss transition metal impurities in covalent materials [4.7]. We shall not develop their full model here, but present a simplified version of it in order to illustrate the importance of screening effects, especially on the level position.

Consider an isolated atom with a n-fold degenerate atomic level E_d, corresponding to n-spin orbitals $\chi_{0\alpha}$, $\alpha = 1,n$ (since we consider the spin explicitly, the occupancy of each $\chi_{0\alpha}$ is zero or unity). We create a defect by coupling this atom to the covalent crystal. The most trivial application is the impurity interstitial. However, the model can represent any other situation (for the vacancy we should couple the defect molecule with the remaining crystal). In a simple Hartree approximation, we write (4.3) for the defect level E_d in the crystal:

$$E_d = E_{d0} + \gamma_{00}(M_0 - Z_0) + \sum_{i \neq 0} \gamma_{0i}(M_i - Z_i) \quad . \tag{4.4}$$

Label 0 corresponds to the defect site and label i to any other atom. We can always write

$$\sum_{i \neq 0} \gamma_{0i}(M_i - Z_i) = \overline{\gamma} \sum_{i \neq 0} (M_i - Z_i) \quad , \tag{4.5}$$

where $\overline{\gamma}$ is the average value of the γ_{0i}.

Consider first the neutral state of the defect for which

$$\sum_{i \neq 0} (M_i - Z_i) = -(M_0 - Z_0) \tag{4.6}$$

and

$$E_d = E_{d0} + J(M_0 - Z_0) \quad , \tag{4.7}$$

where

$$J = \gamma_{00} - \overline{\gamma} \tag{4.8}$$

is the effective Coulomb energy at the defect site. When there are q (q > 0 or < 0) excess electrons, they produce a long-range potential qe^2/r which reduced to $qe^2/\varepsilon r$ because it polarizes the medium. We can therefore write, using simple electrostatics,

$$\sum_{i \neq 0} (M_i - Z_i) = -(M_0 - Z_0) + \frac{q}{\varepsilon} \quad , \tag{4.9}$$

but because the excess charge $-qe/\varepsilon$ is usually very small (in silicon and germanium $\varepsilon \simeq 10$) we neglect it and still apply (4.7).

Let us discuss qualitatively the effects of the coupling between the impurity states and the crystal eigenstates. Before coupling, the atomic wave functions are exact eigenstates of energy E_d. If E_d falls within the gap the situation corresponds to the limit of completely localized states. The introduction of the coupling between each defect state and the crystal states will then produce two effects.

a) On the localized state: As discussed in Sect.3.1, the energy of the local state, if it still exists, will shift from E_d to E_L. The associated wave function will be delocalized and have components on the perfect crystal eigenstates.

b) On the band states: For energies falling within the allowed bands of the unperturbed system, the coupling with the defect leads to wave functions which are linear combinations of the unperturbed band states and the defect states.

To calculate M_0 we must use (4.2),

$$M_0 = \sum_k f(k) \sum_{\alpha=1}^{n} |a_{0\alpha}^k|^2 \quad .$$ (4.10)

For a given impurity wave function $\chi_{0\alpha}$, the quantity $\sum_k |a_{0\alpha}^k|^2$ can be divided into three parts, corresponding to the contributions of the valence band (VB), the localized states (L) and the conduction band (CB). We define these three contributions as

$$\theta_v = \sum_{k(VB)} |a_{0\alpha}^k|^2$$ (4.11)

$$\theta_L = |a_{0\alpha}^L|^2$$ (4.12)

$$\theta_c = \sum_{k(CB)} |a_{0\alpha}^k|^2 \quad .$$ (4.13)

In the general case, these quantities depend on α, but for simplicity we neglect this dependence assuming that θ_v, θ_L, θ_c represent some average over α. An important point is that the sum of these three quantities is equal to unity.

We can now express M_0. The contribution of the valence band to M_0 is equal to $n\theta_v$ since all states in the valence band are filled and all the impurity states with $\alpha = 1,n$ contribute. The contribution of the local states (if they exist) is $(Z_0 + q)\theta_L$ since there are only $(Z_0 + q)$ electrons available. We then obtain

$$M_0 = n\theta_v + (Z_0 + q)\theta_L \quad .$$ (4.14)

The quantities θ_v and θ_L are functions of E_d and also of the coupling strength, but there is no need to specify these functions for a qualitative treatment. Self-consistency can be obtained only if there is a localized state since we must fill this state with $(Z_0 + q)$ electrons.

The detailed behavior of θ_v, θ_L, θ_c versus E_d depends upon the numerical parameters. However, there are general features that do not depend on these particular values. For instance, when E_d falls deep within the valence band, the impurity states only couple with the states of this particular band and θ_v becomes equal to unity, while θ_L and θ_c tend to zero. On the other hand, when E_d lies at high energies within the conduction band, then θ_c is unity and θ_v and θ_L are zero. The schematic behavior of θ_v and θ_c are given in Fig. 4.1, where E_v and E_c indicate the valence-band maximum and the conduction-band minimum, respectively. Physically, the interesting situation corresponds to cases where E_d lies in the energy range of the band gap. Then we

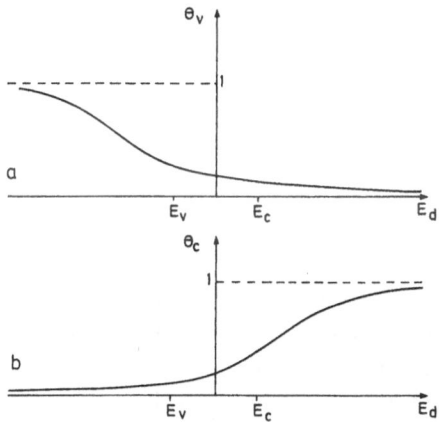

Fig.4.1a,b. Variation with the position of the impurity level of the contributions of the valence band (a) and of the conduction band (b) to the charge of the impurity site

use for θ_v and θ_c a linear interpolation. Considering for simplicity a symmetrical situation where $\theta_v(E_d) = \theta_c(-E_d)$, we can write

$$\theta_v \simeq a - b \frac{E_d}{E_g} \,, \tag{4.15}$$

and

$$\theta_c \simeq a + b \frac{E_d}{E_g} \,, \quad \text{with} \quad a > 0, \quad b > 0 \,, \tag{4.16}$$

where $E_g = E_c - E_v$ is the band-gap width. The number M_0 of electrons on the impurity atom will be, using (4.14),

$$M_0 = n\left(a - b\frac{E_d}{E_g}\right) + (Z_0 + q)(1 - 2a) \quad . \tag{4.17}$$

We are interested mainly in the dependence of M_0 on the charge state of the defect, corresponding to changes of q by integer values. In our linearized treatment this variation is represented by the quantity

$$\frac{dM_0}{dq} = -\frac{nb}{E_g}\frac{dE_d}{dq} + 1 - 2a \quad . \tag{4.18}$$

Since dE_d/dq is related by (4.7) to dM_0/dq, we obtain

$$\frac{dE_d}{dq} = J\frac{dM_0}{dq} \quad . \tag{4.19}$$

Solving both equations we find

$$\frac{dM_0}{dq} = \frac{1-2a}{1+\dfrac{nbJ}{E_g}} \quad . \tag{4.20}$$

The values of the parameters a and b can be obtained from the Green's functions theory of Chap.3. Here, we shall use estimates corresponding to the physical situations of interest (transition-metal impurities, vacancies, etc.). In these situations there is strong coupling between the defect and the band states. Therefore, the bound-state wave functions must be substantially delocalized. We take $\theta_L \simeq 0.6$ (this is the value obtained in Chap.3 for the vacancy). As θ_L is equal to $1 - \theta_v - \theta_c$, (4.15,16) indicate that a $\simeq 0.2$; $\theta_v(E_c)$ must lie between this value and zero, so that we take $\theta_v(E_c) \simeq 0.1$, which gives $b \simeq 0.2$ from (4.16). The other important parameter is the effective Coulomb term J equal to $\gamma_{00} - \bar{\gamma}$. If γ_{00} is calculated from the free-atom wave functions, it is equal to 20 to 30 eV for transition metals, 10 eV for silicon, and 19 eV for diamond (see Sect.4.2.2c). The quantity $\bar{\gamma}$ is less easy to determine, but most of it will correspond to a charge transfer with the nearest neighbors. It is then simply equal to e^2/R, i.e., 9 eV for diamond and 5 eV for silicon. In all cases J will then take values ranging between 5 and 20 eV. We choose the value J = 10 eV. This choice is not essential since it will be shown in the following that the results turn out to be independent of J. The number of defect states n is ten for transition-metal atoms, six for the vacancy (corresponding to the threefold-degenerate T_2 state with up or down spins). Since E_g is of the order of 1 eV, the quantity nb J/E_g is much greater than one (about 20 in our example with $n = 10$). We have therefore

$$\frac{dM_0}{dq} \cong \frac{E_g}{nbJ} (1 - 2a) \quad . \tag{4.21}$$

Using for a the value of 0.2 given before, we find that dM_0/dq takes a value (0.03) very small compared to unity. The important conclusion is that a change of one unit in the charge state of the defect only leads to a very small change in the electron population at the defect site.

The charge-state change induces a variation of E_d. From equations (4.19-21) we obtain

$$\frac{dE_d}{dq} \cong \frac{E_g}{nb} (1 - 2a) \quad . \tag{4.22}$$

This expression, which does not depend at all upon the Coulomb term J, allows us to calculate the derivative dE_L/dq of the localized state energy. For this we use the formalism of chapter 3 which gives E_L as the solution of expression (3.30). Taking the derivative of this expression with respect to q, and using (3.31) which gives the value of θ_L, we obtain the general expression

$$\frac{dE_L}{dq} = \theta_L \frac{dE_d}{dq} \quad , \tag{4.23}$$

that is,

$$\frac{dE_L}{dq} = \frac{E_g}{nb} (1 - 2a)^2 \quad . \tag{4.24}$$

In the numerical example considered above, this quantity is equal to 0.18 E_g, which means that up to 5 different charge states can give levels within the gap. This conclusion that several charge states of a defect can give rise to several localized levels is verified experimentally: four charge states of the vacancy are known to exist in the forbidden gap (in silicon); most of the transition elements can exhibit more than two charge states. This is confirmed by recent, more elaborate calculations: a) in the local-density pseudopotential formalism [4.8], where dE_L/dq is found to be 0.25 eV for the vacancy and b) in the empirical tight-binding method [4.9] when the assumption of the local neutrality leads to dE_L/dq of the order of 0.21 eV.

4.1.3 The CNDO Method

The complete neglect of differential overlap (CNDO) approximation [4.10] is
in many respects similar to the self-consistent tight-binding technique described previously. It is based on a semiempirical representation of the
Hartree-Fock operator, and overlap integrals are neglected as in the tight-
binding approximation. The Hamiltonian is separated into two parts: the core
part formed by the kinetic energy term plus the potential due to the bare
nuclei; and the dielectronic part, which contains the repulsions between the
valence electrons. The corresponding matrix elements are obtained from semiempirical rules. The basis functions are free-atom Slater orbitals. Usually
a minimal basis set, i.e., one s and three p orbitals per atom, is considered.

The matrix elements corresponding to the core part are determined differently when they are intra or interatomic. The intraatomic terms are taken to
be diagonal in the natural s-p basis and are written as

$$(H_c)_{\alpha i, \alpha i} \simeq E_\alpha - \sum_j (Z_j - 1/2)\gamma_{ij} \quad . \tag{4.25}$$

Here α stands for the orbital index s or p, i and j are the atomic indices
and the γ_{ij} are Coulomb terms, Z_j being the charge of the bare nucleus j.
The energy E_α is given by the average of the ionization potential and electron affinity. The interatomic matrix elements of H_c are taken to be

$$(H_c)_{\alpha i, \alpha j} \simeq \frac{1}{2} (\beta_i + \beta_j) K S_{\alpha i, \beta j} \quad , \tag{4.26}$$

where β_i and β_j are empirically chosen parameters and $S_{\alpha i, \alpha j}$ is the overlap
integral $\langle \phi_{\alpha i} | \phi_{\beta j} \rangle$ between the two Slater functions $\phi_{\alpha i}$ and $\phi_{\beta j}$. The intra-
atomic matrix elements of the electron-electron operator H_e are written as

$$(H_e)_{\alpha i, \alpha i} = -\frac{1}{2} M_{\alpha i}\gamma_{ii} + \sum_j M_j \gamma_{ij} \quad , \tag{4.27}$$

where $M_{\alpha i}$ is the electron population of the orbital $\phi_{\alpha i}$, while M_j is the
total electron population of atom j. All other interatomic elements of H_e
are given by the following rule:

$$\int \phi_{\alpha i}^*(\underline{r})\phi_{\beta j}(\underline{r}) \frac{e^2}{|\underline{r}-\underline{r}'|} \phi_{\gamma l}^*(\underline{r}')\phi_{\delta m}(\underline{r}')d\tau d\tau' = \delta_{\alpha i, \beta j}\delta_{\gamma l, \delta m}\gamma_{il} \quad . \tag{4.28}$$

The empirical parameters of the method are β_i, E_s, E_p, and also the exponent in the Slater function. Their values are determined from a best fit to
known physical properties. The validity of such a method then depends on its

ability to predict more properties than those which have been used to determine the parameters. This has been done in the past for small molecules, with
recent applications to diamond and silicon. The parameters for small molecules
and extended systems are found to differ considerably. For instance, $-\beta$ decreases from 21 eV to about 10 eV in diamond [4.11]. The same tendency is
found for $-E_S$, which decreases from 14 eV to 7 eV. Suitable parameters can be
found to reproduce at least the valence-band structure and the equilibrium
distance, but not the compressibility.

This method has been applied to the vacancy in diamond, in a cluster calculation [4.12]. The neutral vacancy is studied in two different ways: (a)
no atomic states at the vacancy site, and (b) atomic orbitals at the vacancy
site with no core. The second method should be preferred in a first-princi-
ples calculation, but its meaning in a semiempirical treatment is not clear.
In both cases the A_1 and T_2 states discussed in Chap.3 are found to fall
within the gap. An attempt to determine the atomic displacements around the
vacancy has also been made. Finally, other charge states of the vacancy in
diamond have been considered, but the results have not been detailed so that
we do not present them here.

4.1.4 The Extended Hückel Theory

We now describe a semiempirical linear combination of atomic orbitals (LCAO)
technique, the first to have been applied to deep levels in semiconductors
[4.13]. The original method is non-self-consistent and based on the following points:

a) Overlap integrals such as $\langle x_{i\alpha}|x_{j\beta}\rangle$ are included as in the full LCAO
   treatments.
b) Intraatomic matrix elements of the Hamiltonian are taken to be

$$\langle x_{i\alpha}|H|x_{i\alpha}\rangle = -I_{i\alpha} \quad , \qquad\qquad (4.29)$$

   where $I_{i\alpha}$ is the ionization energy of the free neutral atom i in the state
   $x_{i\alpha}$.
c) Interatomic matrix elements of H are written as

$$\langle x_{i\alpha}|H|x_{j\beta}\rangle = -\frac{K}{2}(I_{i\alpha} + I_{j\beta})\langle x_{i\alpha}|x_{j\beta}\rangle \quad , \qquad\qquad (4.30)$$

   where K is an empirical constant adjusted to fit experiments as closely
   as possible (for diamond 1.75 seems a "good" value).

Once these expressions are adopted, the technique consists simply to solve the secular equation $\det|H - ES| = 0$ (S being the overlap matrix) to obtain the eigenvalues (i.e., the one-electron energies) and the eigenfunction But because the overlap integrals only decrease slowly in real space, this remains a heavy numerical procedure, and for systems lacking translational symmetry (as is the case for defects) the number of atoms must be restricted to a cluster of limited size. However, in any case the problem is much simpler than a first principles LCAO calculation of the same size.

An important point concerning such methods is that they should provide good band structures for the perfect crystal before applying them to defect systems. It has been shown [4.14] that, starting from free-atom Slater orbitals, this is only roughly true for diamond. However, a slight modification of these orbitals (corresponding to a 25% increase of the exponent of the s function) improves the situation considerably, leading to a very good band structure. We then must hope that one-electron energies corresponding to defects in such systems are correctly predicted.

Extended Hückel theory can be applied to the prediction of equilibrium geometries. For this, the total energy is calculated as a sum of the individual energies of all occupied one-electron states and then this quantity is minimized with respect to the atomic configuration. This procedure was found to be successful for many molecules [4.15], correctly predicting bond angles. However, great care must be taken in its application because it cannot be thought to lead to an exact value for the total energy, the corrections due to the difference of nuclear-nuclear and electron-electron repulsions being, in general, substantial.

Improvements on this technique must include self-consistency (i.e., charge-transfer effects) and the corrections to the total energy we have just mentioned above. Such refinements have been made recently in a manner which will now be described. The first improvement concerns the prediction of the one-electron energies, i.e., the definition of the Hamiltonian matrix elements in the LCAO basis, to which we must add terms depending of the electron-charge distribution. For this atomic charges, as defined by MULLIKEN [4.16] are calculated starting from the expression of the electron density at point $\underline{r}$, $n(\underline{r})$ in terms of atomic functions,

$$n(\underline{r}) = 2 \sum_k f(k) \sum_{ij,\alpha\beta} a_{i\alpha}^{*k} a_{j\beta}^{k} \chi_{i\alpha}^{*}(\underline{r}) \chi_{j\beta}(\underline{r}) \quad , \tag{4.31}$$

the $a_{i\alpha}^{k}$ being defined in Sect.4.1 as the coefficients of the expansion of the kth eigenfunction ψ_k. The integral (4.31) can be partitioned in an infinite number of ways between the orbitals. One possible choice is to attribute to the orbital $\chi_{i\alpha}$, the partial contribution $n_{i\alpha}(\underline{r})$,

$$n_{i\alpha}(\underline{r}) = 2 \text{ Re} \left\{ \sum_k f(k) \sum_{j\beta} a_{i\alpha}^{*k} a_{j\beta}^k \chi_{i\alpha}^*(\underline{r}) \chi_{j\beta}(\underline{r}) \right\} , \qquad (4.32)$$

whose integral over space gives the number of electrons M_i attributed to the orbital $\chi_{i\alpha}$,

$$M_i = 2 \text{ Re} \left\{ \sum_k f(k) \sum_{j\beta} a_{i\alpha}^{*k} a_{j\beta}^k S_{i\alpha,j\beta} \right\} , \qquad (4.33)$$

where S is the overlap matrix. From this, we can choose [4.17] to write the diagonal terms of the Hamiltonian matrix under a form analogous to the tight-binding formula (4.3)

$$\langle \chi_{i\alpha} | H | \chi_{i\alpha} \rangle = - I_{i\alpha} + \sum_{j\beta} (M_{j\beta} - Z_{j\beta}) \gamma_{i\alpha,j\beta} , \qquad (4.34)$$

where $Z_{j\beta}$ is the population of $\chi_{j\beta}$ in the neutral atom. The relation (4.34) gives the usual expression for neutral atoms, but allows for effects due to charge redistribution (we do not detail here the form used in [4.17] for the $\gamma_{i\alpha,j\beta}$). Then the total energy E_T is estimated [4.18] as

$$E_T = \sum_i \varepsilon_i + R_N - I_{ee} , \qquad (4.35)$$

where the first term on the right corresponds to the sum of the one-electron energies, the second one to the nuclear-nuclear repulsion term, the third one correcting for the electron-electron interaction term which was counted twice. Such treatment is fundamentally much more correct than the usual evaluation of E_T in extended Hückel theory.

4.1.5 Self-Consistent Green's Function Calculations

We shall now describe two recent calculations of the electronic states of the neutral vacancy which are quite close in spirit and lead to comparable results. Both are based on the local density formalism [4.2,19] and use a Green's function technique to solve the vacancy problem in a self-consistent manner. Both exploit the fact that, because the self-consistent perturbation potential of the neutral vacancy is of very short range, a Green's function approach can be used.

Let us first describe shortly the principles of one of these calculations [4.20,21]. Equations (3.65,66) of Chap.3 are used to express the wave function ψ of the perturbed system in terms of ψ_0, the unperturbed wave function, and of V, the perturbative potential. Equation (3.65), for instance, is re-written under the form

$$V\psi = V\psi_0 + VG_0V\psi \quad . \tag{4.36}$$

In cases where the perturbation V is strongly localized, any function $V\psi$ is also localized in space and an expansion of this expression in terms of few localized orbitals can be rapidly convergent. This procedure is, in this case, more advantageous than a direct expansion of ψ in terms of localized orbitals. We thus write

$$V(\underline{r})\psi(\underline{r}) \simeq \sum_i c_i V(\underline{r})\varphi_i(\underline{r}) \quad , \tag{4.37}$$

where the $\varphi_i(\underline{r})$ are the localized orbitals (in principle, not orthogonal) and the c_i are unknown coefficients which must be determined so that the right-hand term in (4.37) is the closest approximation to $V\psi$. This is best done using a variational procedure, which leads to the following set of equations [4.20,21]

$$\sum_j \langle\varphi_i|(I - VG_0)V|\varphi_j\rangle c_j = \langle\varphi_i|V|\psi_0\rangle \quad . \tag{4.38}$$

For bound states we take $\psi_0 = 0$, and we obtain an homogeneous set of equations which only have solutions at energies E such that

$$\det\left\{\langle\varphi_i|(I - VG_0)V|\varphi_j\rangle\right\} = 0 \quad , \tag{4.39}$$

while, for energies E lying within the allowed bands, we have to solve (4.39) by matrix inversion. To determine $V\psi$ from (4.38) or (4.39) and then ψ, it is necessary to represent G_0 under matrix form. G_0 is diagonal in the basis formed by the Bloch functions ψ_0 of the perfect crystal. For numerical convenience any ψ_0 is expressed as a combination of a second set of localized orbitals $\phi_i(\underline{r})$ that repeats periodically in each unit cell and whose coefficients are determined by a least-square fit to ψ_0. Once this is done, it is trivial to express any matrix element $\langle\underline{r}|G_0|\underline{r}'\rangle$ (defined in Chap.3) under the form

$$\langle\underline{r}|G_0|\underline{r}'\rangle = \sum_{\psi_0} \langle\underline{r}|\psi_0\rangle \langle\psi_0|G_0|\psi_0\rangle \langle\psi_0|\underline{r}'\rangle \quad , \tag{4.40}$$

that is

$$<\underline{r}|G_0|\underline{r}'> = \sum_{m,n} \Phi_m^*(\underline{r})\left(G_0\right)_{m,n} \Phi_n(\underline{r}') \quad , \tag{4.41}$$

which allows a straighforward solution of (4.38) and (4.39), while $\psi(r)$ is expressed in terms of the second set of orbitals $\Phi_m(\underline{r})$, the product $V\psi$ being expressed in terms of the first set $\varphi_i(\underline{r})$.

The calculations are performed in a self-consistent manner, using a pseudo-potential method within a local density scheme. From this point of view it is similar to the work of [4.22], which however had the disadvantage of being a supercell calculation (i.e., introducing a periodic array of vacancies with finite interactions between them). The Green's function calculation is supe-rior, leading in principle to the exact result if the expansions discussed above are complete enough. Both calculations agree in finding that the self-consistent $V(\underline{r})$ corresponding to the neutral vacancy is strongly localized in space, thereby justifying the whole procedure used. It is also found to be practically of spherical symmetry, the angular anisotropy being smaller than $\pm 10\%$.

The results for the change in the density of states or in the total number of states are completely similar to those obtained from the tight-binding methods of Chap.3, and we need not recall them here. Let us simply mention that the T_2 bound state is now found at $+0.7$ eV (i.e., higher in the gap than in the tight-binding calculations of the references quoted in Sect.3.2.7 [3.25,26]) while the A_1 resonance is at -1.1 eV. All these results concern the undistorted neutral vacancy, all atoms remaining at their perfect crystal positions. The effect of relaxations and distorsions are not taken into ac-count.

Another calculation completely similar to the one discussed above is de-scribed in [4.25]. However, the mathematical techniques involved in the cal-culation are simpler. Use is made of the fact that the central quantities to be computed have the property of being localized near the defect site (this should not be true for all defects, but is at least correct for the neutral vacancy). For instance, as shown in Chap.3, a key quantity when calculating the change in density of states is $\det|I - G_0V|$. We have seen that the dimen-sionality of this determinant and of the perturbation V are the same. If V is strongly localized, then a small basis set of functions localized in the neighborhood of the defect site will provide the correct answer. This is also true for the change $\delta n(\underline{r})$ in the electron density which, in a self-consistent calculation, will have the same localization as V itself. This change can be expressed in terms of Green's functions [using (3.54,57)]. This leads to

$$\delta n(\underline{r}) = - \frac{2}{\pi} \operatorname{Im} \left\{ \int f(E)[G(\underline{r},\underline{r}) - G_0(\underline{r},\underline{r})]dE \right\} , \tag{4.42}$$

the summation being taken over the occupied states for which f(E) is unity. Using Dyson's equation (3.74), this expression can be rewritten

$$\delta n(\underline{r}) = \frac{2}{\pi} \operatorname{Im} \left\{ \int f(E)[G_0 - (I - G_0 V)^{-1} G_0]dE \right\} . \tag{4.43}$$

The calculation of these localized quantities is made, using a LCAO technique (with ten orbitals per atom). Symmetrized orbitals are formed on each shell of atoms surrounding the defect site (shell orbitals). These shell orbitals are not orthogonal, but orthogonalized shell orbitals are built by orthogonal-izing each shell orbital to the orbitals on all the shells closer to the de-fect site. This is done for the first three shells of neighbors surrounding the vacancy (a total of 28 atoms). As usual the results are iterated up to self-consistency, i.e., the calculated density $\delta n(\underline{r})$ gives the potential $V(\underline{r})$ which leads to an identical $\delta n(\underline{r})$ through (4.43). The results are equi-valent to those of [4.20,21] except for minor differences. Another calculation in the same spirit has been attempted [4.26], but there, the potential is not in agreement with the above cited references, and it is not clear that in this case self-consistency has been reached.

Another important result of these calculations concerns the variation of the localized state energy with the charge state of the defect. It is found that this state increases by 0.25 eV when going from V^+ to V^0 or from V^0 to V^-, which is the expected order of magnitude as discussed in Sect.4.1.2.

4.1.6 Thomas-Fermi Interpretation of the Self-Consistent Potential

A very simple method to find a quite good approximation to the self-consistent potential $V(\underline{r})$ discussed above for the neutral vacancy is provided by the Thomas-Fermi method. While it seems at first sight only applicable to high-density metals, we shall see that it is in fact quite well adapted to the covalent semiconductors, which we discuss here.

The most simple version, under which self-consistency effects induced by a perturbation can be discussed, is linear screening. This assumes a per-turbative potential $V(\underline{r})$ weak enough so that its effect can be treated to first order. $V(\underline{r})$ induces a change in electron density $\delta n(\underline{r})$ linear in V, which itself gives rise to a self-consistent potential δV linear in δn; this potential δV adds to the bare perturbation V_b to finally give the total per-turbative potential $V(\underline{r})$. We can therefore write formally

$$V_b = \varepsilon V \quad .$$

(4.44)

Equation (4.44), written under matrix form, defines the dielectric matrix ε that can be expressed either in real or reciprocal space, leading respectively to the following expressions:

$$V_b(\underline{r}) = \int \varepsilon(\underline{r},\underline{r}')V(\underline{r}')d^3\underline{r}'$$

(4.45)

$$V_b(\underline{q} + \underline{K}) = \sum_{\underline{K}'} \varepsilon(\underline{q} + \underline{K}, \underline{q} + \underline{K}')V(\underline{q} + \underline{K}') \quad .$$

(4.46)

In the reciprocal space $\underline{q}$ belongs to the first Brillouin zone, and $\underline{K}$, $\underline{K}'$ are reciprocal lattice vectors. The simplified form of ε in the reciprocal space, where only matrix elements differing by a reciprocal lattice vector do not vanish, results from the translational periodicity of the perfect crystal. The most important part of ε is formed by its diagonal elements, the non-diagonal elements giving rise to the "local field" effects. We have thus plotted in Fig.4.2 the quantity $\varepsilon(\underline{q},\underline{q})$ calculated in [4.27] from Penn's model (it only depends on q, the modulus of $\underline{q}$, in view of the isotropy of the model). On the same figure is plotted an approximate form for this quantity, written as

$$\varepsilon(\underline{q},\underline{q})_{\big|} = \frac{1 + \dfrac{\lambda^2}{q^2}}{1 + \dfrac{1}{\varepsilon(0)}\dfrac{\lambda^2}{q^2}} \quad ,$$

(4.47)

where we introduce the experimental value for $\varepsilon(0)$, and where λ is taken to be the Thomas-Fermi wave number of the metal having the same average density. It is clear that (4.47) gives an extremely good approximation to the calculated curve.

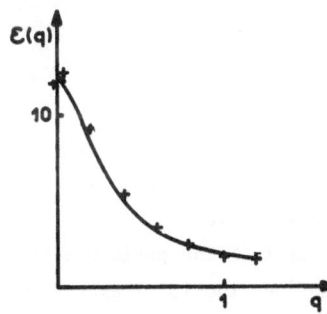

Fig.4.2. Variation of the dielectric constant versus wave vector. ($\times$): from [4.27]; (——) calculated from (4.47)

Let us now investigate the physics contained in (4.47) in order to be able to generalize it. For this, it is convenient to rewrite it under the form

$$\frac{1}{\varepsilon(q,q)} = \left[1 - \frac{1}{\varepsilon(0)}\right] \frac{1}{1 + \frac{\lambda^2}{q^2}} + \frac{1}{\varepsilon(0)} \quad . \tag{4.48}$$

Applied to a perturbative bare potential which behaves as $1/r$, this leads to

$$V(r) = \frac{1}{\varepsilon(0)r} + \left[1 - \frac{1}{\varepsilon(0)}\right] \frac{e^{-\lambda r}}{r} \quad . \tag{4.49}$$

This means that a part, equal to $\varepsilon(0)^{-1}$ of the long-range Coulomb potential, remains unscreened, while the remaining part $[1 - \varepsilon(0)^{-1}]$ becomes completely screened exactly as it would be in the metal having the same electron density treated in the Thomas-Fermi approximation. Furthermore, we can easily check that these densities are high enough for a Thomas-Fermi treatment to become valid. All of this explains the success of (4.47) as well as its physical meaning.

To generalize this treatment to the neutral vacancy case is not a diffi-cult matter. A neutral defect does not present any long-range Coulombic bare potential, so that it is completely screened, and it is enough to apply di-rectly the Thomas-Fermi approximation as in a metal. To compare with the Green's functions results of [4.21] and [4.25], we take the same bare per-turbative potential $V_b(r)$ (the opposite of the atomic bare pseudopotential) and solve the following set of equations:

$$\delta n(\underline{r}) = \frac{1}{3\pi^2} \left(\frac{2m}{\hbar^2}\right)^{3/2} \left[(E_F - V_0 - V)^{3/2} - (E_F - V_0)^{3/2}\right] \tag{4.50}$$

$$\nabla^2 V(\underline{r}) = -4\pi e^2 \delta n(\underline{r}) + \nabla^2 [V_{xc}(\underline{r}) + V_b(\underline{r})] \tag{4.51}$$

$$V_{xc}(\underline{r}) = -\alpha \frac{3}{2\pi} (3\pi^2)^{1/3} e^2 \left[(n_0 + \delta n)^{1/3} - n_0^{1/3}\right] \quad . \tag{4.52}$$

Equation (4.50) defines the Thomas-Fermi approximation where the change in electron density $\delta n(\underline{r})$ is expressed in a local form of the potential V_0 in the perfect crystal and of the total perturbative potential V. Equation (4.51) is Poisson's equation relating the electrostatic part of the perturbative po-tential $V - (V_{xc} + V_b)$ to $\delta n(\underline{r})$. Equation (4.52) defines the local approximation $V_{xc}(\underline{r})$ used for the exchange-correlation potential used in all these calcula-tions.

We solve these equations, first by linearizing them in $V(\underline{r})$, and second in a fully numerical manner as described in [4.28]. Figure 4.3 shows both results compared to the exact spherical average V(r) of the Green's functions calculations. Two conclusions emerge.

a) The Thomas-Fermi treatment leads to a practically exact result.
b) The linear version of the Thomas-Fermi approximation is practically as accurate as the full treatment.

This means that, for deep levels which arise from neutral defects, the self-consistent potential is practically identical to what would occur in the metal of the same density and is well-approximated by a linearized Thomas-Fermi treatment. This last treatment has the advantage of giving an analytical value for the screening length. In [4.28] V(r) is found to have an exponential screening in exp(-γr) (with γ equal to 1.46 for silicon), which explains why the perturbation is so efficiently screened in space (in a radius of 4 a.u., i.e., smaller than the nearest neighbor's distance 4.44 a.u.).

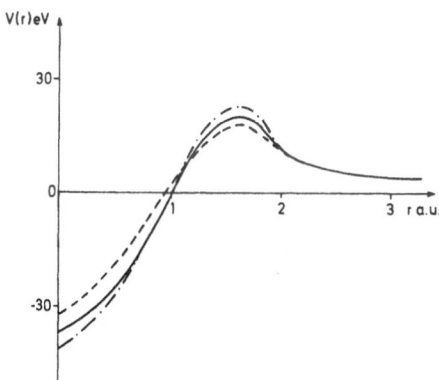

Fig.4.3. Total perturbation potential versus distance. (———) Thomas-Fermi approximation; (---): exact calculation from [4.21]; (—·—·—) linearized Thomas-Fermi approximation

4.2 Many-Electron Effects. The Configuration Interaction

In this second part we analyze the importance of correlation effects for the most simple defect, the vacancy. This is done within the defect molecule model (built from the four dangling orbitals, which point towards the vacancy site), using the configuration interaction technique. Since this will be a complex matter, we first present the theory on the simplest system possible, i.e., the H_2 molecule, where we introduce all notations and discussions which will be useful for the case of the vacancy.

4.2.1 The H_2 Molecule

We shall first present the notations that will be used throughout this section and then derive the complete solution for the problem. Finally, we shall discuss different possible one-electron approximations as well as their range of validity.

a) Notations

The full Hamiltonian H of the H_2 molecule can be written as

$$H = h_1 + h_2 + e^2/r_{12} \quad , \tag{4.53}$$

where 1 and 2 index the two electrons, e^2/r_{12} is the interelectronic repulsion and h_1 and h_2 are one-electron operators containing kinetic and potential energy terms:

$$h = - \frac{\hbar^2}{2m} \Delta - \frac{e^2}{r_A} - \frac{e^2}{r_B} \quad , \tag{4.54}$$

r_A and r_B being the distances between an electron and the two (A and B) H nuclei (Fig.4.4).

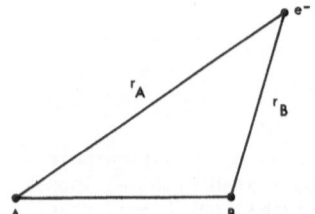

Fig.4.4. The hydrogen molecule

To solve this many-electron problem, it is usual to start with a given set of one-electron spin orbitals (product of an orbital part by a spin part) which we designate by $\varphi_\alpha(\underline{r})$ (α including orbital and spin indices).

Then, all possible independent products of two of these spin orbitls, such as $\varphi_\alpha(\underline{r}_1) \cdot \varphi_\beta(\underline{r}_2)$, are considered, and all possible linear combinations of such products, which are antisymmetric in the permutation of the two electrons, are taken. This is completely equivalent to building the Slater determinants,

$$|\varphi_\alpha \varphi_\beta> = \frac{1}{\sqrt{2}} \begin{vmatrix} \varphi_\alpha(\underline{r}_1) & \varphi_\beta(\underline{r}_1) \\ \varphi_\alpha(\underline{r}_2) & \varphi_\beta(\underline{r}_2) \end{vmatrix} \quad . \tag{4.55}$$

This set of determinants is then considered as a basis for the total Hamiltonian H. Usually the one-electron orbitals are determined from a variational principle leading to the Hartree-Fock equations.

Here, we take as basic one-electron orbitals the free-atom orbitals $\varphi_{a\uparrow}$, $\varphi_{a\downarrow}$, $\varphi_{b\uparrow}$, $\varphi_{b\downarrow}$, the second index specifying the spin direction corresponding to the ground state (1s) of atoms A and B. In general, they are not orthogonal, i.e., their overlap $S = \langle\varphi_a|\varphi_b\rangle$ is nonzero. Since this introduces complications in the treatment we prefer to define the bonding and antibonding orbitals ψ_B and ψ_A given by

$$\psi_{\substack{B\\A}} = \frac{\varphi_a \pm \varphi_b}{\sqrt{2(1\pm S)}} \quad , \tag{4.56}$$

which lead to four orthonormal spin orbitals $\psi_{B\uparrow}$, $\psi_{B\downarrow}$, $\psi_{A\uparrow}$, $\psi_{A\downarrow}$. We first determine the one-electron matrix elements

$$\langle\psi_{B\uparrow}|h|\psi_{B\uparrow}\rangle = \langle\psi_{B\downarrow}|h|\psi_{B\downarrow}\rangle = \overline{E} - \frac{\Delta}{2}$$

$$\langle\psi_{A\uparrow}|h|\psi_{A\uparrow}\rangle = \langle\psi_{A\downarrow}|h|\psi_{A\downarrow}\rangle = \overline{E} + \frac{\Delta}{2} \quad , \tag{4.57}$$

all other matrix elements vanishing because of spatial or spin symmetry. We also need matrix elements of two-electron terms, i.e., e^2/r_{12}, which are of the general form

$$\langle\varphi_\alpha\varphi_\beta|\varphi_\gamma\varphi_\delta\rangle = \int \varphi_\alpha^*(1)\varphi_\beta(1) \frac{e^2}{r_{12}} \varphi_\gamma^*(2)\varphi_\delta(2)d\tau_1 d\tau_2 \quad . \tag{4.58}$$

We must notice first that such elements vanish when the spin part of α and β and also of γ and δ are different. When they are identical, integration over spins gives unity. We shall therefore use (4.58) in terms of the orbital part alone.

Instead of evaluating completely these integrals, we use Mulliken's approximation which introduces only little error and allows a reduction of the problem to two fundamental parameters, as we shall see later. This approximation replaces any product of orbital parts $\varphi_a(\underline{r})\varphi_b(\underline{r})$ by the quantity $S[|\varphi_a(\underline{r})|^2 + |\varphi_b(\underline{r})|^2]/2$. In terms of bonding orbitals this readily leads to the nonvanishing terms

$$\langle\psi_B\psi_B|\psi_B\psi_B\rangle = \langle\psi_A\psi_A|\psi_A\psi_A\rangle = \langle\psi_B\psi_B|\psi_A\psi_A\rangle = \overline{J} \tag{4.59}$$

and

$$\langle\psi_B\psi_A|\psi_B\psi_A\rangle = \frac{U}{2} \quad , \tag{4.60}$$

where the two parameters $\bar{J}$ and $U/2$ are defined by the elementary Coulomb integrals,

$$J = \frac{\langle\varphi_a\varphi_a|\varphi_a\varphi_a\rangle + \langle\varphi_a\varphi_a|\varphi_b\varphi_b\rangle}{2} \qquad (4.61)$$

and

$$U = \frac{\langle\varphi_a\varphi_a|\varphi_a\varphi_a\rangle - \langle\varphi_a\varphi_a|\varphi_b\varphi_b\rangle}{\sqrt{1-S^2}} \;. \qquad (4.62)$$

To second order in S, the problem becomes equivalent to the neglect of all overlap terms, a point which we shall use later.

It is trivial to show that each diagonal matrix element of H in the basis of the Slater determinants built from $\psi_{B\uparrow}$, $\psi_{B\downarrow}$, $\psi_{A\uparrow}$, $\psi_{A\downarrow}$ will contain the term $2\bar{E}+\bar{J}$, which we thus take as the origin of the energy. The model contains only two parameters, a one-electron term Δ and a Coulomb integral U, in terms of which all differences in energy can be expressed.

b) Exact Solution

We can calculate all matrix elements between Slater determinants, either directly in this case, or using general rules given by Slater [4.29]. The Hamiltonian matrix in the basis $|\psi_{B\uparrow}\psi_{B\downarrow}\rangle$, $|\psi_{B\uparrow}\psi_{A\downarrow}\rangle$, $|\psi_{B\downarrow}\psi_{A\uparrow}\rangle$, $|\psi_{B\uparrow}\psi_{A\uparrow}\rangle$, $|\psi_{B\downarrow}\psi_{A\downarrow}\rangle$ and $|\psi_{A\uparrow}\psi_{A\downarrow}\rangle$ is

$$\begin{bmatrix}
-\Delta & 0 & 0 & 0 & 0 & \frac{U}{2} \\
0 & 0 & -\frac{U}{2} & 0 & 0 & 0 \\
0 & -\frac{U}{2} & 0 & 0 & 0 & 0 \\
0 & 0 & 0 & -\frac{U}{2} & 0 & 0 \\
0 & 0 & 0 & 0 & -\frac{U}{2} & 0 \\
\frac{U}{2} & 0 & 0 & 0 & 0 & +\Delta
\end{bmatrix} \;. \qquad (4.63)$$

Its solutions are given in Table 4.1.

Table 4.1. Characteristics of the eigenstates for the H_2 molecule

Energy	Wave function	Spin	Symmetry
$\pm \sqrt{\Delta^2 + U^2/4}$	combination of $\|\psi_{B\uparrow}\psi_{B\downarrow}\rangle$ and $\|\psi_{A\uparrow}\psi_{A\downarrow}\rangle$	$S = 0$ singlet	$^1\Sigma_g$ symmetric
$-\dfrac{U}{2}$	$\|\psi_{A\uparrow}\psi_{B\uparrow}\rangle$ $\dfrac{\|\psi_{A\uparrow}\psi_{B\downarrow}\rangle + \|\psi_{A\downarrow}\psi_{B\uparrow}\rangle}{\sqrt{2}}$ $\|\psi_{A\downarrow}\psi_{B\downarrow}\rangle$	$S = 1$ triplet	$^3\Sigma_g$ symmetric
$+\dfrac{U}{2}$	$\dfrac{\|\psi_{A\uparrow}\psi_{B\downarrow}\rangle - \|\psi_{A\downarrow}\psi_{B\uparrow}\rangle}{\sqrt{2}}$	$S = 0$ singlet	$^1\Sigma_u$ antisymmetric

These solutions are represented in Fig.4.5, where the quantity 4E/U (E being the energy) is plotted versus Δ/U.

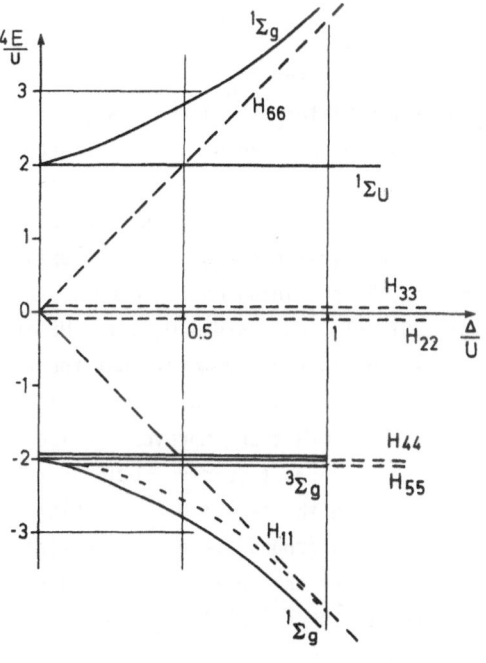

Fig.4.5. Energy states (reduced to the Coulomb integral U) of the H_2 molecule as a function of the strength of the correlation. The dashed lines give the behavior of the diagonal terms which correspond to the use of the Hartree-Fock approximation. The dotted line is the ground state for the unrestricted Hartree-Fock solution. The H_{ii} are the diagonal elements of matrix (4.63)

c) Restricted Hartree-Fock Solution and Configuration Interaction

In the Hartree-Fock (HF) approximation the spin orbitals of the ground-state Slater determinant are solutions of the one-electron equation:

$$\left(h + \sum_k <\psi_k\psi_k>\right)\psi_i(\underline{r}) - \sum_k <\psi_k\psi_i>\psi_k(\underline{r}) = \varepsilon_i\psi_i(\underline{r}) \quad , \tag{4.64}$$

where $<\psi_k\psi_i>$ is a short-hand notation for $\int \psi_k^*(\underline{r}')\psi_i(\underline{r}')e^2/|\underline{r} - \underline{r}'|d\tau'$ and the ψ_k, ψ_i are occupied one-electron spin orbitals. As before, we search for one-electron solutions which are linear combinations of the free atom states. However, even in that case the HF equation (4.64) can have different solutions corresponding to the "restricted HF approximation" or the "unrestricted" one.

In the restricted solution it is assumed that the HF Hamiltonian has the fu symmetry of the molecule. Thus, its solutions must be either symmetric (the bonding states $\psi_{B\uparrow}$ or $\psi_{B\downarrow}$) or antisymmetric (the antibonding states $\psi_{A\uparrow}$ and $\psi_{A\downarrow}$). The states of lowest energy clearly turn up to be the bonding states so that the ground-state Slater determinant is $|\psi_{B\uparrow}\psi_{B\downarrow}>$ whose energy is simply given by the corresponding diagonal term H_{11} of (4.63). Excited states will be Slater determinants obtained by replacing $\psi_{B\uparrow}$ or $\psi_{B\downarrow}$ or both in $|\psi_{B\uparrow}\psi_{B\downarrow}>$. The energies of all these HF excited states correspond to the other H_{ii} terms and are plotted as dashed lines on Fig.4.5. We can group theses states into configurations as in free atoms. The ground state configuration is obviously ψ_B^2, i.e. with two electrons in ψ_B corresponding to $|\psi_{B\uparrow}\psi_{B\downarrow}>$. The first excited configuration can be labelled $\psi_B^1\psi_A^1$ corresponding to the four possible slater determinants $|\psi_{B\uparrow}\psi_{A\uparrow}>$, $|\psi_{B\uparrow}\psi_{A\downarrow}>$, $|\psi_{B\downarrow}\psi_{A\uparrow}>$ and $|\psi_{B\downarrow}\psi_{A\uparrow}>$. The last excited configuration is evidently ψ_A^2 with the Slater determinant $|\psi_{A\uparrow}\psi_{A\downarrow}>$.

Different degrees of sophistication can thus be imagined. We first consider the pure HF solutions, whose levels are given in Fig.4.6 in the case $\Delta/U > 0.5$. Then we can solve the full Hamiltonian, but only within each different configuration, which may be called a "first order or an intraconfiguration" treatment. Finally, allowance for the remaining matrix elements gives the interaction between configurations. Clearly, when $\Delta/U > 0.5$, the HF approximation already gives the ground state of correct symmetry, but the excitation energies are poorly described (two doublets instead of a singlet and a triplet). A first-order configuration interaction improves considerably the situation leading practically to the correct spectrum but underestimating the transition energies. However, in the case $\Delta/U < 0.5$, Fig.4.5 shows that the HF approximation gives the wrong ground state. Even a first-order configuration interaction does not improve this point. We are then in the strongly correlated limit and a restricted HF calculation is of no

significance. Fortunately, it is possible to go beyond this approximation
and use an "unrestricted" form for the HF solution, which we now describe.

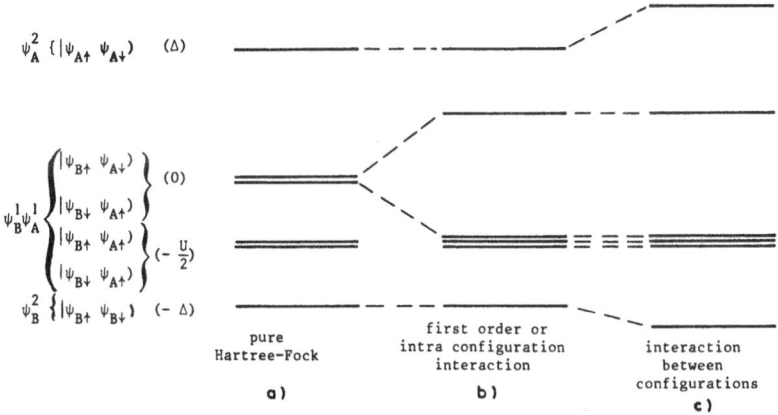

Fig.4.6a-c. Energy levels of the H_2 molecule for $\Delta/U > 0.5$ (a) pure Hartree-Fock solutions; (b) first-order or intraconfiguration interaction; (c) interaction between configurations

d) Unrestricted Hartree-Fock Solution

Here, we do not impose any more symmetric solutions to (4.64). Let us then
try a ground-state determinant of the form $|\psi_{1\uparrow}\psi_{2\downarrow}>$, where the orbital parts
ψ_1 and ψ_2 are allowed to be different. To simplify we neglect the overlap
integral S and write the normalized $\psi_{1\uparrow}$ and $\psi_{2\downarrow}$ under the form

$$\psi_{1\uparrow} = \sqrt{1/2 + n_\uparrow}\ \varphi_{a\uparrow} + \sqrt{1/2 - n_\uparrow}\ \varphi_{b\uparrow} \tag{4.65}$$

$$\psi_{2\downarrow} = \sqrt{1/2 + n_\downarrow}\ \varphi_{a\downarrow} + \sqrt{1/2 - n_\downarrow}\ \varphi_{b\downarrow} \quad . \tag{4.66}$$

However, we impose the physical constraint that the final charge density must
be symmetrical so that the total number of electrons on atom A be 1 as well
as on atom B. This clearly leads to

$$n_\uparrow + n_\downarrow = 0 \quad , \tag{4.67}$$

leaving only one undetermined parameter, which then has to be obtained from
the solution of the Hartree-Fock equation (note at this stage that the 're-stricted' form corresponds to $n_\uparrow$ and $n_\downarrow$ equal to zero and contains no para-meter). We thus write (4.64) corresponding to the Slater determinant $|\psi_{1\uparrow}\psi_{2\downarrow}>$

(the exchange term vanishes because the term $\langle\psi_{1\uparrow}\psi_{2\downarrow}\rangle$ is zero after integration over spins). The equations corresponding to the orbital parts ψ_1 and ψ_2 are, respectively,

$$(h + \langle\psi_2\psi_2\rangle)\psi_1(\underline{r}) = \varepsilon_1\psi_1(\underline{r}) \quad , \quad \text{and} \tag{4.68}$$

$$(h + \langle\psi_1\psi_1\rangle)\psi_2(\underline{r}) = \varepsilon_2\psi_2(\underline{r}) \quad . \tag{4.69}$$

When expressing ψ_1, ψ_2 given by (4.65,66) in the atomic basis φ_a and φ_b, (4.68,69) can be written under matrix form with different 2×2 matrices $h_\uparrow$ for spins up and $h_\downarrow$ for spins down. We can express them in terms of the parameters $\overline{E}$, Δ, $\overline{J}$, and U defined above by (4.57,61,62):

$$h_\uparrow = \overline{E} + \overline{J} + \begin{bmatrix} n_\downarrow U & -\frac{\Delta}{2} \\ -\frac{\Delta}{2} & -n_\downarrow U \end{bmatrix} \quad \text{and} \tag{4.70}$$

$$h_\downarrow = \overline{E} + \overline{J} + \begin{bmatrix} n_\uparrow U & -\frac{\Delta}{2} \\ -\frac{\Delta}{2} & -n_\uparrow U \end{bmatrix} \quad . \tag{4.71}$$

The corresponding eigenvalues are

$$\varepsilon_\uparrow = \overline{E} + \overline{J} \pm \sqrt{(n_\downarrow U)^2 + \Delta^2/4} \tag{4.72}$$

$$\varepsilon_\downarrow = \overline{E} + \overline{J} \pm \sqrt{(n_\uparrow U)^2 + \Delta^2/4} \quad . \tag{4.73}$$

In both equations we have to retain only the lowest eigenvalue. The calculation of the corresponding eigenfunctions is straightforward and, when expressed under the form (4.65) and (4.66), gives for $n_\uparrow$ and $n_\downarrow$ the following expressions:

$$n_\uparrow = - \frac{Un_\downarrow}{\sqrt{(Un_\downarrow)^2 + \Delta^2/4}} \tag{4.74}$$

$$n_\downarrow = - \frac{Un_\uparrow}{\sqrt{(Un_\uparrow)^2 + \Delta^2/4}} \tag{4.75}$$

Because $n_\uparrow$ is equal to $-n_\downarrow$, this set of equations reduces to one equation with one unknown, giving for $n_\uparrow$

$$n_\uparrow = \sqrt{1-\Delta^2/U^2}/2 \quad . \tag{4.76}$$

This equation has solutions only when $\Delta/U < 1$, i.e., in the limit where the Coulomb term is predominant. From the variational principle this solution must be lower in energy than the restricted solution because the wave function is more flexible. At the value $\Delta/U = 1$, $n_\uparrow$ and $n_\downarrow$ vanish and the unrestricted solution becomes identical with the restricted one. If we calculate the ground state energy

$$E_0 = (\psi_{1\uparrow}\psi_{2\downarrow}|H|\psi_{1\uparrow}\psi_{2\downarrow}) = <\psi_1|h|\psi_1> + <\psi_2|h|\psi_2> + <\psi_1\psi_1|\psi_2\psi_2> \quad , \tag{4.77}$$

we obtain

$$E_0 = 2\overline{E} + \overline{J} - \frac{U}{2} - \frac{\Delta^2}{2U} \quad , \tag{4.78}$$

where the value of $n_\uparrow$ given by (4.76) has been injected. This energy is plotted as a dotted line in Fig.4.5, which shows that a much better value, tending towards the exact value at $\Delta/U \to 0$ (i.e., for an infinite separation) is obtained in this approximation.

It is also interesting to know how this situation can be improved by allowing for the interaction between configurations. From the determinant $|\psi_{1\uparrow}\psi_{2\downarrow}>$, we first deduce $|\psi_{1\downarrow}\psi_{2\uparrow}>$, which evidently has the same energy. We can also build $|\psi_{1\uparrow}\psi_{2\uparrow}>$ and $|\psi_{1\downarrow}\psi_{2\downarrow}>$, which will become degenerate with the two previous ones when $\Delta/U \to 0$. Finally, the highest determinants $|\psi_{1\uparrow}\psi_{1\downarrow}>$ and $|\psi_{2\uparrow}\psi_{2\downarrow}>$ form a twofold-degenerate level much higher in energy.

A final point can be made concerning the shape of the one-electron orbitals. When $n_\uparrow$ is taken to be positive, $\psi_{1\uparrow}$ is localized mainly on atom A and $\psi_{2\downarrow}$ on atom B as shown on Fig.4.7. In the limit $\Delta/U \to 0$ the determinant $|\psi_{1\uparrow}\psi_{2\downarrow}>$ becomes $|\varphi_{a\uparrow}\varphi_{b\downarrow}>$, the two electrons gaining more energy when they are on the two different atoms.

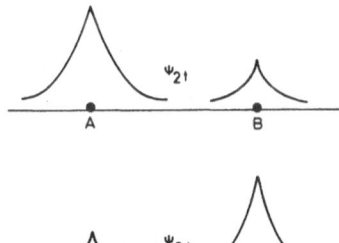

Fig.4.7 Localization in space of the spin orbitals of an H_2 molecule in the unrestricted Hartree-Fock approximation

e) Direct Use of Symmetry Properties

We shall now describe how the results of the simple restricted Hartree-Fock theory can be extended by direct use of molecular symmetry properties, resulting in a much better prediction for the excitation spectrum without the need for further calculations.

For this, let us remember that ψ_B and ψ_A are respectively symmetric and antisymmetric. They are said to be basis functions of the Σ_g and Σ_u irreducible representations of the symmetry point group C_{2v}, using the language of group theory (see Sect.1.4). To build Slater determinants, we first have to form products of these spin orbitals. The spatial part, $\psi_B(1)\psi_B(2)$ behaves as $\Sigma_g \times \Sigma_g$, and is obviously symmetric (thus $\Sigma_g \times \Sigma_g = \Sigma_g$). The same is true for $\psi_A(1) \cdot \psi_A(2)$ ($\Sigma_u \times \Sigma_u = \Sigma_g$). On the other hand, $\psi_A(1) \cdot \psi_B(2)$ or $\psi_A(2) \cdot \psi_B(1)$ are antisymmetric ($\Sigma_g \times \Sigma_u = \Sigma_u$). To simplify the diagonalization of the whole 6×6 Hamiltonian matrix, we can use the following considerations:

a) H commutes with the symmetry operations. The eigenstates can be labelled by their symmetry properties; here, they must belong either to Σ_g or to Σ_u.

b) Only states that are basis functions of the same row of the same degenerate irreducible representation can interact together; here only symmetric Σ_g states interact with other Σ_g states and not with Σ_u states.

c) H also commutes with the total spin, so that we can also classify the states with respect to their spin quantum numbers.

Let us detail this for the H_2 molecule. The Hartree-Fock ground state is $|\psi_{B\uparrow}\psi_{B\downarrow}>$. It is a combination of products whose orbital part is formed by $\psi_B(1)\psi_B(2)$, which then has a Σ_g symmetry. It is an eigenstate of the total spin with $S = 0$ and $S_z = 0$. It can thus be written as a $^1\Sigma_g$ state (the superscript corresponds to $2S + 1$). Let us now form the first excited configuration B^1A^1. This configuration contains the four determinants $|\psi_{B\uparrow}\psi_{A\uparrow}>$, $|\psi_{B\downarrow}\psi_{A\downarrow}>$, $|\psi_{B\uparrow}\psi_{A\downarrow}>$ and $|\psi_{B\downarrow}\psi_{A\uparrow}>$. All are combinations of simple products whose orbital part contains $\psi_B(1) \cdot \psi_A(2)$ or $\psi_B(2) \cdot \psi_A(1)$ and have the symmetry of $\Sigma_g \times \Sigma_u = \Sigma_u$, i.e., are antisymmetric. They are thus classified with respect to orbital symmetry. Now it is clear that $|\psi_{B\uparrow}\psi_{A\uparrow}>$ and $|\psi_{B\downarrow}\psi_{A\downarrow}>$ are total spin eigenstates ($S = 1$, $S_z = \pm 1$), which is not the case for $|\psi_{B\uparrow}\psi_{A\downarrow}>$ and $|\psi_{B\downarrow}\psi_{A\uparrow}>$. However, standard theorems on the addition of spins allow us to say that $[|\psi_{B\uparrow}\psi_{A\downarrow}> + |\psi_{B\downarrow}\psi_{A\uparrow}>]/\sqrt{2}$ is the $S_z = 0$ component of the triplet (with $S = 1$), while the other combination $[|\psi_{B\uparrow}\psi_{A\downarrow}> - |\psi_{B\downarrow}\psi_{A\uparrow}>]/\sqrt{2}$ has spin $S = 0$, $S_z = 0$. Finally, the last configuration $|\psi_{A\uparrow}\psi_{A\downarrow}>$ behaves as $\Sigma_u \times \Sigma_u = \Sigma_g$ with spin $S = 0$. The results are summarized in Table 4.2.

Table 4.2. Classification of the eigenstates of the H_2 molecule according to their configuration

Configuration	Eigenstate	Symmetry
B^2	$\lvert \psi_{B\uparrow} \psi_{B\downarrow} \rangle$	$^1\Sigma_g$
$B^1 A^1$	$\lvert \psi_{B\uparrow} \psi_{A\uparrow} \rangle$	
	$\dfrac{\lvert \psi_{B\uparrow} \psi_{A\downarrow} \rangle + \lvert \psi_{B\downarrow} \psi_{A\uparrow} \rangle}{\sqrt{2}}$	$^3\Sigma_u$
	$\lvert \psi_{B\downarrow} \psi_{A\downarrow} \rangle$	
	$\dfrac{\lvert \psi_{B\uparrow} \psi_{A\downarrow} \rangle - \lvert \psi_{B\downarrow} \psi_{A\uparrow} \rangle}{\sqrt{2}}$	$^1\Sigma_u$
A^2	$\lvert \psi_{A\uparrow} \psi_{A\downarrow} \rangle$	$^1\Sigma_g$

The important thing to notice is that, simply by using symmetry arguments, we can derive wave functions that are better than the simple Slater determinants without doing any calculation. It is immediately apparent that the use of the diagonal terms of H in the basis given in Table 4.2 gives exactly the same results as the first-order configuration interaction case of Fig.4.6; it gives the correct shape of the excitation spectrum when $\Delta/U > 0.5$, i.e., provides the major correction in the weakly correlated limit. To improve these results still further, we have simply to allow for configuration interaction between states of the same symmetry (i.e., here only between $\lvert \psi_{B\uparrow} \psi_{B\downarrow} \rangle$ and $\lvert \psi_{A\uparrow} \psi_{A\downarrow} \rangle$ which both belong to $^1\Sigma_g$).

4.2.2 The Vacancy in Covalent Systems

One important conclusion derived from WATKINS' work [4.30] is that the simple one electron picture of the vacancy discussed in Chap.3 can give a quite satisfactory interpretation of the experimental results. This seems true not only for the isolated vacancy, but also for many vacancy associated defects. Such a conclusion is, at first sight, in contradiction with the available many-electron calculations which tend to conclude that correlation effects are always important and that a one electron model should not be a good approximation. This is why we intend to discuss here in detail such effects and show how it is possible to reconcile the two points of view.

We develop here, for the vacancy, a theory similar to what we have done in Sect.4.2.1 for the H_2 molecule. In fact, it was by this sort of calculation that the theory of defects in diamond started. This was first done by COULSON and KEARSLEY [4.31] with the precise aim of interpreting the GR 1 optical absorption band observed in irradiated diamond [4.32]. Later on, important information about vacancies in silicon were obtained by WATKINS from electron paramagnetic resonance experiments, concluding that large Jahn-Teller distorsions occur near such defects. We shall consider Jahn-Teller effects separately [Ref.1.1, Chap.8] and assume in the following that the atoms remain at their perfect crystal positions.

a) Description of the Model

We want to discuss the many-electron effects that arise at the localized states of the vacancy. Since it is impossible to evaluate them exactly, we use the defect-molecule approximation discussed in detail in Chap.3, whose main features are briefly recalled.

As shown in Fig.4.8, the defect-molecule model considers the four dangling sp^3 orbitals $x_{i0}(i = 1$ to $4)$ to form an isolated subsystem, treated as a molecule with $n = 4 - q$ electrons, where q is the qth charge state of the vacancy, denoted V^q (for instance, V^0 is the neutral state, V^+ the positive charge state, V^- the negative one ...). The full electronic Hamiltonian of this small molecule is

$$H = \sum_{i=1}^{n} h_i + \frac{1}{2} \sum_{i=1}^{n} \sum_{j=1}^{n} \frac{e^2}{r_{ij}} \quad , \qquad (4.79)$$

where h_i are equivalent one-electron operators equal to the sum of the kinetic energy of the electron plus its potential energy in the field of the four nuclei. This potential also includes a part due to the outer electrons. The second term represents the repulsive interaction of the vacancy electrons. The problem is thus quite similar to the H_2 molecule except that we have four nuclei and n electrons. Here we shall consider the cases n = 3, 4, 5 (i.e., q = +1, 0, -1).

We have thus to specify which one-electron spin orbitals we want to start with. Their number is obviously eight (they are $x_{i0\uparrow}$, $x_{i0\downarrow}$ with i = 1 to 4).

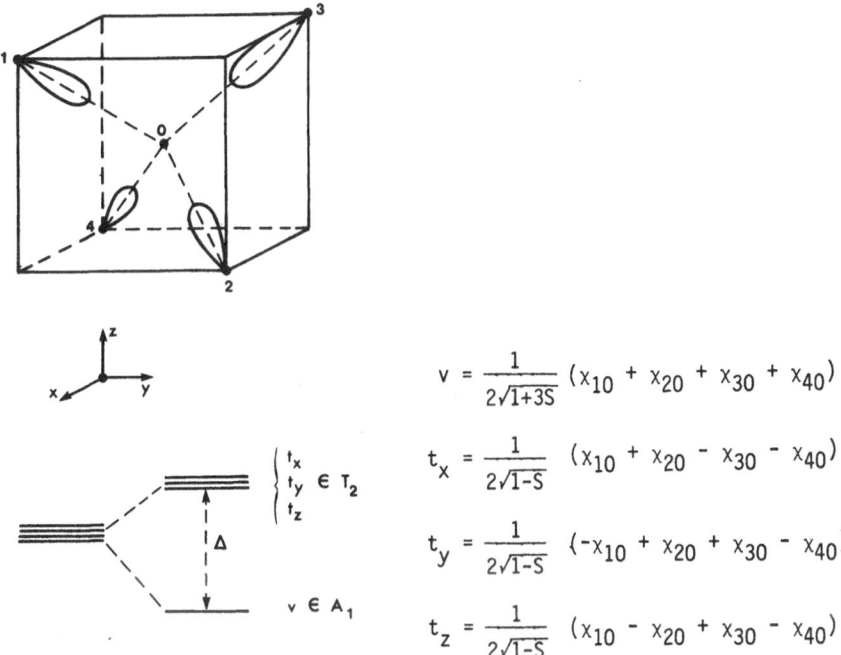

$$v = \frac{1}{2\sqrt{1+3S}} (x_{10} + x_{20} + x_{30} + x_{40})$$

$$t_x = \frac{1}{2\sqrt{1-S}} (x_{10} + x_{20} - x_{30} - x_{40})$$

$$t_y = \frac{1}{2\sqrt{1-S}} (-x_{10} + x_{20} + x_{30} - x_{40})$$

$$t_z = \frac{1}{2\sqrt{1-S}} (x_{10} - x_{20} + x_{30} - x_{40})$$

Fig.4.8. The defect-molecule model for the vacancy and corresponding one-electron energy states. The orbitals v, t_x, t_y, t_z are combinations of the four hybrid orbitals which have the symmetry of the molecule

However, these atomic sp_3 hybrids, which point from atom i towards the vacancy site, do not form an orthonormal set, a fact which would complicate the calculation. So, we form four symmetric combinations of these hybrids, whose orbital parts are given in Fig.4.8, where S is the overlap between two sp_3 hybrids, i.e., $\langle x_{i0}|x_{j0}\rangle$. We have already discussed the advantage of these orbitals v, t_x, t_y, and t_z in our discussion of the tight-binding treatment in Chap.3. (They automatically diagonalize any operator having the full symmetry of the system.) We thus write for the matrix elements of the one-electron operators

$$\langle v|h|v\rangle = \bar{E} - \frac{3\Delta}{4} \tag{4.80}$$

and

$$\langle t_\alpha|h|t_\alpha\rangle = \bar{E} + \frac{\Delta}{4} \quad , \quad \alpha = x,y,z \quad , \tag{4.81}$$

which, exactly as for H_2, depend only upon two parameters, the average energy $\bar{E}$ and the splitting Δ of the one-electron states. Of course, the completely symmetric orbitals corresponds to the lowest energy, analogous to the bonding orbital of H_2.

We now consider the interelectronic terms of the form $\langle \alpha\beta | \gamma\delta \rangle$ as in (4.58). We again use Mulliken's approximation, writing

$$\langle \alpha\beta | \gamma\delta \rangle = \frac{1}{4} S_{\alpha\beta} S_{\gamma\delta} \langle \alpha^2 + \beta^2 | \gamma^2 + \delta^2 \rangle \quad , \tag{4.82}$$

where $S_{\alpha\beta}$ and $S_{\gamma\delta}$ are the overlap matrix elements. With this approximation, all electron-electron terms can be reduced to two fundamental integrals J and J' defined by

$$J = \langle x_{i0} x_{i0} | x_{i0} x_{i0} \rangle, \quad i = 1 \text{ to } 4 \tag{4.83}$$

and

$$J' = \langle x_{i0} x_{i0} | x_{j0} x_{j0} \rangle, \quad i \neq j = 1 \text{ to } 4 \quad . \tag{4.84}$$

In terms of the symmetry orbitals v, t_x, t_y, and t_z, we obtain the following nonvanishing independent terms:

$$\left. \begin{array}{l} \langle vv | vv \rangle \\ \langle t_\alpha t_\alpha | t_\alpha t_\alpha \rangle \\ \langle vv | t_\alpha t_\alpha \rangle \\ \langle t_\alpha t_\alpha | t_\beta t_\beta \rangle \end{array} \right\} = \frac{J + 3J'}{4} = \overline{J}$$

$$\langle vt_\alpha | vt_\alpha \rangle = \frac{(1+S)^2}{(1+3S)(1-S)} \cdot \frac{U}{4} \tag{4.85}$$

$$\langle t_\alpha t_\beta | t_\alpha t_\beta \rangle = \frac{U}{4}$$

$$\langle vt_y | t_x t_z \rangle = - \frac{1+S}{\sqrt{(1+3S)(1-S)}} \cdot \frac{U}{4} \quad \text{with}$$

$$U = J - J' \quad . \tag{4.86}$$

The bielectronic integrals given by (4.85) correspond in the same order to the C, D, E, F, G, J, K integrals of COULSON and KEARSLEY [4.31], their A and B integrals corresponding to $\langle v|h|v \rangle$ and $\langle t_x|h|t_x \rangle$. The advantage of our formulation is that the integrals are reduced to two fundamental terms, $\overline{J}$ and U. Furthermore, exactly as for H_2, we see that the overlap S can be neglected to second order since $S \simeq 0.1$, when calculated with the free-atom orbitals.

Our model contains four parameters $\overline{E}$, Δ, $\overline{J}$, and U. However, it is clear that the average repulsion of two electrons is $(J + 3J')/4$ and thus, for a system with n electrons, the term $n\overline{E} + [n(n-1)/2]\,\overline{J}$ will appear in all levels.

We can therefore take this energy as an origin, and all the excitation energies will only depend on Δ and U, i.e., on two parameters, exactly as for H_2.
This important simplification will allow us to do a systematic discussion of
the results versus the relative strength of correlation effects, measured by
Δ/U.

b) The Different Configurations

We start from the Hartree-Fock ground state. We consider the restricted
Hartree-Fock solution, where the one-electron Hamiltonians are assumed to
be completely symmetric. Thus, the one-electron eigenfunctions can only be
the symmetric orbitals $v_\uparrow$, $v_\downarrow$, $t_{x\uparrow}$, $t_{x\downarrow}$, $t_{y\uparrow}$, $t_{y\downarrow}$, $t_{z\uparrow}$, and $t_{z\downarrow}$. Since v has
the lower energy, the Hartree-Fock ground state will always correspond to
filling the A_1 state with two electrons and the T_2 state with the remaining
n-2 electrons. The ground-state configuration is in all cases $v^2 t^{n-2}$, using
an obvious notation.

To perform the complete treatment, as was done for the H_2 molecule, we must
first write down the ground-state Slater determinant. For the positive vacancy
V^+, n = 3, one determinant is $|v_\uparrow v_\downarrow t_{x\uparrow}\rangle$, but clearly there are many other
possibilities $|v_\uparrow v_\downarrow t_{x\downarrow}\rangle, |v_\uparrow v_\downarrow t_{y\uparrow}\rangle$, etc., all having the same energy. A simi-
lar situation will occur for the first excited configuration $v\,t^2$, where
$|v_\uparrow t_{x\uparrow} t_{x\downarrow}\rangle$, $|v_\downarrow t_{x\uparrow} t_{x\downarrow}\rangle$, etc., are also degenerate. Then, simple Hartree-Fock
theory will result in a few states having strong degeneracy and will certain-
ly not be very accurate. It is thus quite necessary to extend the method described for H_2 where symmetry properties have been used to derive more accurate wave functions and energies.

The general procedure is relatively simple. Suppose we want to analyse
a configuration $v^p\,t^{n-p}$, where p takes the values 0, 1, or 2. We know the
orbital v is the basis function of the A_1 nondegenerate representation of
the group T_d, while t denotes one of the three basis functions t_x, t_y, or
t_z of the threefold-degenerate representation T_2. The symmetry of the orbital
part of any wave function belonging to $v^p\,t^{n-p}$ will thus belong to the representation $(A_1 \times A_1 ...) \times (T_2 \times T_2 ...)$. Standard group theory allows the decomposition of such products as linear combinations of irreducible representations in
a systematical manner and to find what combinations of the Slater determinants
have the correct symmetry properties (see Sect.4.1). Once this is done, it
is still necessary to classify the states as eigenstates of the total spin S,
eliminating those which cannot satisfy the Pauli principe (for instance, a
determinant containing $t_x t_x t_x$ as orbital part is not acceptable). The whole

procedure is detailed in a paper by MULLIKEN [4.33], which gives detailed rules for many cases of interest.

We do not intend to give the details of such standard techniques, but will only give the wave functions for the different configurations of the V^+, V^0, and V^- charge states. This is done in Table 4.3 where we also give the diagonal term of H, i.e., the average value of H for the given wave function. For any degenerate state we only give the first wave function and, for simplicity, we write the Slater determinants under the form $|...\alpha\bar{\beta}...>$, where α means a spin orbital α with spin up, $\bar{\alpha}$ with spin down. We also note x for t_x, y for t_y, and z for t_z.

The results quoted in Table 4.3 correspond exactly to what we called an intraconfiguration or first-order configuration interaction, where the exact Hamiltonian matrix is diagonalized separately within each configuration. We can see that this problem is automatically solved by symmetry because each representation of a given spin occurs once and only once within each configuration (remember that only states belonging to the same irreductible representation can interact via the Hamiltonian).

It is important to know the changes which are brought by the inclusion of configuration interaction. Only states of the same symmetry interact and Table 4.3 shows that this leads at most to the diagonalization of 3×3 matrices (this is the case, for instance, for the $^2T_2$ of V^+ which occurs in the v^2t, vt^2, and t^3 configurations). The matrix elements of H between these states (given in Table 4.3) can be expressed as combinations of matrix elements between orthonormal Slater determinants and calculated in terms of the parameters Δ and U using the rules derived by SLATER [4.29]. We give these matrices in Table 4.4 in the case where their dimension is 2×2 or 3×3 (for 1×1 the solution of Table 4.3 is exact). We adopt the convention that the ordering of the states corresponds to increasing p in the configurations $v^p t^{n-p}$. The diagonal terms are evidently those of Table 4.3.

The results of this complete diagonalization, including configuration interaction, are given in Fig.4.9, where 4E/U is plotted versus the dimensionless parameter Δ/U, which is a measure of the relative strength of the electron-electron interactions. Only the lowest states are given. The dashed lines of the right side represent the results of the intraconfiguration interaction treatment corresponding to the energies given in Table 4.3; the dashed lines drawn in regions belonging to the left hand side of each figure correspond to the first-order expansion of the energies in Δ/U, i.e., to the strongly correlated limit. The two sets of dashed lines intersect in the cross-hatched regions.

c) Discussion of the Results

We shall first give a simple picture of the two limits $\Delta/U \to 0$ and $\Delta/U \to \infty$ and then compare our predictions with other calculations based on the same model, but using more complex forms for the electron-electron interactions. Finally, we shall give estimates for Δ and U, trying to answer the question of the relative importance of correlation effects.

The strongly correlated limit $\Delta/U \to 0$ is easy to understand. We have n electrons to put on the four atoms (or into eight spin orbitals $|x_{i0}{}^{\uparrow}\rangle$ or $|x_{i0}{}^{\downarrow}\rangle$). If Δ is negligible, the electrons tend to lower their energy by sitting on different atoms. The ground state is obtained for a maximum number of electron pairs on different neighbors. The degeneracy of this ground state is strong ($2^3 C_4^3 = 32$ for V^+, $2^4 = 16$ for V^0, $2^3 C_4^3 = 32$ for V^-). Higher excited states correspond to new pairing of electrons on the same atomic orbital and thus lie at $+U$ above the ground state for V^+, $+U$ and $+2U$ for V^0 and V^-. When Δ/U increases from this limit, there is linear splitting in the ground-state manifold only for V^+ and V^- due to the fact that, only in these cases, can electron or hole hopping occur (this is impossible for V^0 because

Table 4.3. Characteristics of the eigenstates for the positive, neutral, and negative states of the vacancy

			Positive vacancy V^+	
Config-uration	State	Orbital degen-eracy	Wave function	Energy
$v^2 t$	$^2T_2$	3	$\|v\bar{v}x\rangle$	$-\frac{5\Delta}{4} - \frac{U}{4}$
vt^2	$^4T_1$	3	$\|vyz\rangle$	$-\frac{\Delta}{4} - \frac{3U}{4}$
	$^2A_1$	1	$\frac{1}{\sqrt{3}}(\|vx\bar{x}\rangle + \|vy\bar{y}\rangle + \|vz\bar{z}\rangle)$	$-\frac{\Delta}{4} + \frac{U}{4}$
	^2E	2	$\frac{1}{\sqrt{6}}(2\|vx\bar{x}\rangle - \|vy\bar{y}\rangle - \|vz\bar{z}\rangle)$	$-\frac{\Delta}{4} - \frac{U}{2}$
	$^2T_1$	3	$\frac{1}{\sqrt{6}}(2\|\bar{v}yz\rangle - \|v\bar{y}z\rangle - \|vy\bar{z}\rangle)$	$-\frac{\Delta}{4}$
	$^2T_2$	3	$\frac{1}{\sqrt{2}}(\|vy\bar{z}\rangle - \|v\bar{y}z\rangle)$	$-\frac{\Delta}{4}$
t^3	$^4A_2$	1	$\|xyz\rangle$	$\frac{3\Delta}{4} - \frac{3U}{4}$
	^2E	2	$\frac{1}{\sqrt{2}}(\|x\bar{y}z\rangle - \|xy\bar{z}\rangle)$	$\frac{3\Delta}{4}$
	$^2T_1$	3	$\frac{1}{\sqrt{2}}(\|xy\bar{y}\rangle - \|xz\bar{z}\rangle)$	$\frac{3\Delta}{4} - \frac{U}{2}$
	$^2T_2$	3	$\frac{1}{\sqrt{2}}(\|xy\bar{y}\rangle + \|xz\bar{z}\rangle)$	$\frac{3\Delta}{4}$

Table 4.3 (continued)

| | | | Neutral vacancy V^0 | |
Config-uration	State	Orbital degen-eracy	Wave function	Energy
v^2t^2	$^1A_1$	1	$\frac{1}{\sqrt{3}}(|v\bar{v}x\bar{x}\rangle + |v\bar{v}y\bar{y}\rangle + |v\bar{v}z\bar{z}\rangle)$	$-\Delta$
	^1E	2	$\frac{1}{\sqrt{6}}(2|v\bar{v}x\bar{x}\rangle - |v\bar{v}y\bar{y}\rangle - |v\bar{v}z\bar{z}\rangle)$	$-\Delta - \frac{3}{4}U$
	$^3T_1$	3	$|v\bar{v}yz\rangle$	$-\Delta - \frac{3}{4}U$
	$^1T_2$	3	$\frac{1}{\sqrt{2}}(|v\bar{v}\bar{y}z\rangle - |v\bar{v}y\bar{z}\rangle)$	$-\Delta - U/4$
vt^3	$^5A_2$	1	$|vxyz\rangle$	$-\frac{3U}{2}$
	$^3A_2$	1	$\frac{1}{2\sqrt{3}}(|v\bar{x}yz\rangle + |vx\bar{y}z\rangle + |vxy\bar{z}\rangle - 3|\bar{v}xyz\rangle)$	$-\frac{U}{2}$
	^3E	2	$\frac{1}{\sqrt{2}}(|vxy\bar{z}\rangle - |vx\bar{y}z\rangle)$	$-\frac{U}{2}$
	$^3T_1$	3	$\frac{1}{\sqrt{2}}(|vz\bar{z}x\rangle - |vy\bar{y}x\rangle)$	$-U$
	$^3T_2$	3	$\frac{1}{\sqrt{2}}(|vz\bar{z}x\rangle + |vy\bar{y}x\rangle)$	$-\frac{U}{2}$
	$^1T_2$	3	$1/2(|vz\bar{z}\bar{x}\rangle + |vy\bar{y}\bar{x}\rangle - |\bar{v}z\bar{z}x\rangle - |\bar{v}y\bar{y}x\rangle)$	-0
	^1E	2	$1/2(|v\bar{x}\bar{y}z\rangle + |\bar{v}xy\bar{z}\rangle - |v\bar{x}y\bar{z}\rangle - |\bar{v}x\bar{y}z\rangle)$	-0
	$^1T_1$	3	$1/2(|vz\bar{z}\bar{x}\rangle - |vy\bar{y}\bar{x}\rangle - |\bar{v}z\bar{z}x\rangle + |\bar{v}y\bar{y}x\rangle)$	$-\frac{U}{2}$
t^4	$^3T_1$	3	$|x\bar{x}yz\rangle$	$\Delta - \frac{3}{4}U$
	$^1A_1$	1	$\frac{1}{\sqrt{3}}(|x\bar{x}y\bar{y}\rangle + |y\bar{y}z\bar{z}\rangle + |z\bar{z}x\bar{x}\rangle)$	Δ
	^1E	2	$\frac{1}{\sqrt{6}}(|x\bar{x}y\bar{y}\rangle + |x\bar{x}z\bar{z}\rangle - 2|y\bar{y}z\bar{z}\rangle)$	$-\frac{3}{4}U$
	$^1T_2$	3	$\frac{1}{\sqrt{2}}(|x\bar{x}\bar{y}z\rangle - |x\bar{x}y\bar{z}\rangle)$	$\Delta - \frac{U}{4}$
			Negative vacancy V^-	
v^2t^3	$^4A_2$	1	$|v\bar{v}xyz\rangle$	$-\frac{3\Delta}{4} - \frac{3U}{2}$
	^2E	2	$\frac{1}{\sqrt{2}}(|v\bar{v}x\bar{y}z\rangle - |v\bar{v}xy\bar{z}\rangle)$	$-\frac{3\Delta}{4} - \frac{3U}{4}$
	$^2T_1$	3	$\frac{1}{\sqrt{2}}(|v\bar{v}xy\bar{y}\rangle - |v\bar{v}xz\bar{z}\rangle)$	$-\frac{3\Delta}{4} - \frac{5U}{4}$
	$^2T_2$	3	$\frac{1}{\sqrt{2}}(|v\bar{v}xy\bar{y}\rangle + |v\bar{v}xz\bar{z}\rangle)$	$-\frac{3\Delta}{4} - \frac{3U}{4}$
vt^4	$^4T_1$	3	$|vx\bar{x}yz\rangle$	$\frac{\Delta}{4} - \frac{3U}{3}$
	$^2A_1$	1	$\frac{1}{\sqrt{3}}(|vx\bar{x}y\bar{y}\rangle + |vy\bar{y}z\bar{z}\rangle + |vz\bar{z}x\bar{x}\rangle)$	$\frac{\Delta}{4} - U/2$
	^2E	2	$\frac{1}{\sqrt{6}}(|vx\bar{x}y\bar{y}\rangle + |vx\bar{x}z\bar{z}\rangle - 2|vy\bar{y}z\bar{z}\rangle)$	$\frac{\Delta}{4} - \frac{5U}{4}$
	$^2T_1$	3	$\frac{1}{\sqrt{6}}(2|\bar{v}x\bar{x}yz\rangle - |vx\bar{y}z\rangle - |vx\bar{x}y\bar{z}\rangle)$	$\frac{\Delta}{4} - \frac{3U}{4}$
	$^2T_2$	3	$\frac{1}{\sqrt{2}}(|vx\bar{x}\bar{y}z\rangle - |vx\bar{x}y\bar{z}\rangle)$	$\frac{\Delta}{4} - \frac{3U}{4}$
t^5	$^2T_2$	3	$|xy\bar{y}z\bar{z}\rangle$	$\frac{5\Delta}{4} - U$

Table 4.4. Matrix elements of the Hamiltonian between the various states of the vacancy

V^+	^2E	$\begin{bmatrix} \frac{\Delta}{4} - \frac{U}{2} & -\frac{\sqrt{3}}{4}U \\[4pt] -\frac{\sqrt{3}}{4}U & \frac{3\Delta}{4} \end{bmatrix}$		V^-	^2E	$\begin{bmatrix} -\frac{3\Delta}{4} - \frac{3}{4}U & -\frac{\sqrt{3}}{4}U \\[4pt] -\frac{\sqrt{3}}{4}U & \frac{\Delta}{4} - \frac{5}{4}U \end{bmatrix}$
	$^2T_1$	$\begin{bmatrix} \frac{\Delta}{4} & \frac{\sqrt{3}}{4}U \\[4pt] \frac{\sqrt{3}}{4}U & \frac{3\Delta}{4} - \frac{U}{2} \end{bmatrix}$			$^2T_1$	$\begin{bmatrix} \frac{3\Delta}{4} - \frac{5}{4}U & -\frac{\sqrt{3}}{4}U \\[4pt] -\frac{\sqrt{3}}{4}U & \frac{\Delta}{4} - \frac{3}{4}U \end{bmatrix}$
	$^2T_2$	$\begin{bmatrix} \frac{5\Delta}{4} - \frac{U}{4} & \frac{\sqrt{2}\,U}{4} & \frac{\sqrt{2}\,U}{4} \\[4pt] \frac{\sqrt{2}\,U}{4} & -\frac{\Delta}{4} & -\frac{U}{4} \\[4pt] \frac{\sqrt{2}\,U}{4} & \frac{U}{4} & \frac{3\Delta}{4} \end{bmatrix}$			$^2T_2$	$\begin{bmatrix} \frac{3\Delta}{4} - \frac{3U}{4} & \frac{U}{4} & \frac{\sqrt{2}}{4}U \\[4pt] \frac{U}{4} & \frac{\Delta}{4} - \frac{3U}{4} & -\frac{\sqrt{2}\,U}{4} \\[4pt] \frac{\sqrt{2}\,U}{4} & -\frac{\sqrt{2}\,U}{4} & \frac{5\Delta}{4} - U \end{bmatrix}$

V^0	$^1A_1$	$\begin{bmatrix} -\Delta & \frac{U}{2} \\[4pt] \frac{U}{2} & +\Delta \end{bmatrix}$
	$^3T_1$	$\begin{bmatrix} -\Delta - \frac{3}{4}U & -\frac{\sqrt{2}}{4}U & U/4 \\[4pt] -\frac{\sqrt{2}}{4}U & -U & \frac{\sqrt{2}}{4}U \\[4pt] \frac{U}{4} & \frac{\sqrt{2}}{4}U & \Delta - \frac{3U}{4} \end{bmatrix}$
	$^1T_2$	$\begin{bmatrix} -\Delta - \frac{U}{4} & -\frac{\sqrt{2}}{4}U & \frac{U}{4} \\[4pt] -\frac{\sqrt{2}}{4}U & 0 & \frac{\sqrt{2}}{4}U \\[4pt] -\frac{U}{4} & \frac{\sqrt{2}\,U}{4} & \Delta - \frac{U}{4} \end{bmatrix}$
	^1E	$\begin{bmatrix} -\Delta - \frac{3}{4}U & -\frac{\sqrt{6}}{4}U & \frac{U}{4} \\[4pt] \frac{\sqrt{6}}{4}U & 0 & \frac{\sqrt{6}}{4}U \\[4pt] \frac{U}{4} & \frac{\sqrt{6}}{4}U & \Delta - \frac{3U}{4} \end{bmatrix}$

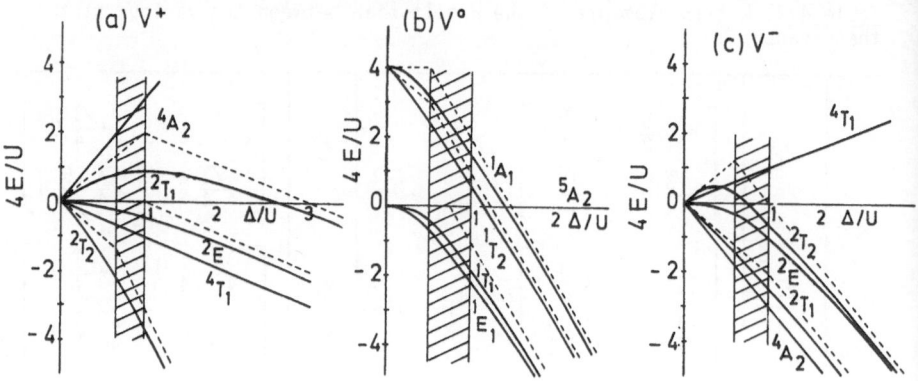

Fig.4.9a-c. Energy levels (reduced to the Coulomb integral U) associated with the vacancy in the molecular model versus the strength the correlation effects: (a) V^+; (b) V^0; (c) V^-. The cross-hatched regions are the frontiers between the strongly and weakly correlated limits (see text)

there is already one electron on each dangling bond). The other limit $\Delta/U \to \infty$ corresponds to the one-electron limit with configurations $v^2 t^{n-2}$, $v t^{n-1}$, and t^n and energies $(n/4 - 2)\Delta$, $(n/4 - 1)\Delta$, and $(n/4)\Delta$, respectively. Again these configurations are highly degenerate but, when U is nonzero, there is a linear splitting of the degenerate manifolds. This leads to shifts of the straight lines which are constants in units of U as shown in Fig.4.9. This represents the intraconfiguration interaction result to this problem. The cross-hatched regions of Fig.4.9 can thus be considered as the frontier between the highly and weakly correlated limits. In all cases these regions lie between $\Delta/U = 1/2$ and 1. Thus for $\Delta/U < 1/2$, we are in the strongly correlated limit, and for $\Delta/U > 1$, the one-electron picture is appropriate if the intraconfiguration interaction is eventually taken into account.

It is now interesting to discuss all "defect-molecule many-electron" calculations in the light of this simple description in order to know what region of our diagram they should correspond to and if the model is really a good representation. Such calculations fall within two groups according to the method used to estimate the one- and two-electron parameters. The first group corresponds to the case where all the two-electron integrals are calculated from given atomic functions (unmodified cases of COULSON and KEARSLEY [4.31] and LARKINS [4.34]). In that case the relations we derived are almost exactly fulfilled and the model gives the correct results in a fairly simple way. It is only for the second group, where the intraatomic Coulomb term is reduced in a semiempirical way, that the model seems to fail and that the ordering of levels becomes sensitive to the detailed values of the parameters. The object of such a reduction was to take some account of the intraatomic correlation effects, and only the term $<x_{i0}x_{i0}|x_{i0}x_{i0}>$ was modified. However, to be consistent with Mulliken's approximation, terms like $<x_{i0}x_{j0}|x_{\ell 0}x_{\ell 0}>$ $(i \neq j \neq \ell)$ should be reduced in the appropriate manner. We thus believe that, even in these modified cases, the model applies, the only change being a reduction of the parameter U. As shown on Fig.4.9 this does not modify the general ordering of the lowest levels but only changes their relative spacing.

Let us now have a look at the detailed numbers which have been used for the parameters. For diamond [4.31,35], the parameter Δ (equal to the $|A| - |B|$ quantity of the authors) takes two values, a calculated one $\Delta \simeq 10 - 12$ eV and a semiempirical one $\Delta \simeq 5$ eV. The same is true for the intraatomic Coulomb integral J for which the calculated value is 19 eV and the semiempirical one $\simeq 13$ eV, while in all cases J' keeps the same value of 8 eV. Thus, in the first case corresponding to calculated values, we have $\Delta \simeq 10$ eV and $U \simeq 11$ eV, leading to a ratio $\Delta/U \simeq 0.9$; in the second one, using the semiempirical values, $\Delta \simeq 5$ eV and $U \simeq 5$ eV, leads to $\Delta/U \simeq 1$, practically the same ratio. The very surprising conclusion is that, for both sets of numbers, we are just at the limit where the restricted Hartree-Fock approximation is a meaningful starting point since configuration interaction within the ground-state manifold is already sufficient to give the correct ordering and even the spacing of the lowest levels. The values for silicon are [4.37]: calculated, $\Delta \simeq 2$ eV, $J \simeq 11.5$ eV, $J' \simeq 5$ eV, and $\Delta/U \simeq 0.32$; empirical, no value is quoted for Δ, but $J = 8.1$ eV, $J' = 5$ eV, (taking again $\Delta \simeq 2$ eV) and $\Delta/U \simeq 0.48$, i.e., we are nearer to the strongly correlated case than for diamond.

It is very interesting to discuss here the results obtained from the use of the generalized valence-bond method [4.36], which has been applied to the neutral vacancy V^0 in diamond and in silicon. Clusters of sixteen atoms C_4H_{12} and Si_4H_{12} were used, for which a complete numerical selfconsistent treatment was performed. Two sets of values, the "modified" and the "unmodified" ones were derived. We shall for the moment only consider the second set of values which corresponds to isolated clusters, not taking into account the effect of the surrounding crystal. This set leads exactly to our predicted ordering of levels in the strongly correlated limit (the numbers correspond to the diagram of Fig.4.9 for V^0 with $\Delta/U \simeq 0.48$ and $U \simeq 8$ eV for carbon and $\Delta/U \simeq 0.4$ and $U \simeq 4.8$ eV for silicon). These results are nearly identical to those of [4.37] for Si, but for diamond the ratio Δ/U is found to be smaller by a factor of two.

d) Validity of the One-Electron Theories

From the above discussion, it appears that current values derived for Δ and U are such that correlation effects are not negligible. From this point of view, only the COULSON and KEARSLEYS' values for diamond are at the limit $\Delta/U \simeq 1$ but, even in that case, intraconfiguration interaction is very important since it introduces a splitting of the order of U (whose smallest value is the semiempirical one $\simeq 5$ eV). This fact was recognized long ago and led many authors to suspect the results obtained by one-electron theories, even if the agreement with experiment was excellent. Thus, we shall now discuss in more detail the values of Δ and U and show that the ratio Δ/U should be increased with respect to the values quoted above.

First, we may suspect some of these values. For instance, the calculated ratio $\Delta(diamond)/\Delta(silicon) \simeq 5$ taken by COULSON et al. is too high. This can be seen from the fact that the contribution of this term to the bulk valence-band width (equal to Δ in a tight-binding calculation) would be of order 10 eV, i.e., about half the total contribution for diamond, which is unrealistic. We choose $\Delta(silicon) = 2$ eV, a value on which all previously discussed calculations agree, and assume with HARRISSON [4.38] that this parameter scales as d^{-2} (where d is the interatomic distance). In this way, the ratio $\Delta(diamond)/\Delta(silicon)$ reduces to 8/3, leading to 5.33 eV for $\Delta(diamond)$.

A second reason for modifying the parameters comes from the meaning of the defect molecule itself. In reality this molecule is coupled to the rest of the crystal. As a result the dangling bond states become delocalized. The one-electron splitting Δ will thus be multiplied by a reducing factor equal to the fractional contribution of the dangling orbitals to the local-state wave function. This factor is about 60% for Si, leading to an effective Δ equal to 1.2 eV, in good agreement with the findings of tight-binding calculations of the $A_1 - T_2$ spacing [4.23]. The corresponding reduction should be less for diamond where the localization is expected to be more important. If we take 80% for this localization, we have Δ (diamond) $\simeq 4.3$ eV.

Another important characteristic of a correct many-electron treatment of the lowest energy states of the vacancy is that interelectronic repulsion should be allowed, not only between electrons belonging to the localized states, but also between these and other electrons belonging to the rest of the crystal. Thus, if we want the "defect-molecule many-electron model to be a useful concept", it is necessary to incorporate this effect as a correction to the values of the electron-electron parameters to be used in the model.

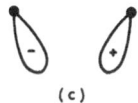

<u>Fig.4.10a-c.</u> Density fluctuations on dangling bonds

This will result in an effective value of U different from the one calculated from the defect molecule alone. The most dramatic change can be simply analyzed as follows. The parameter U is equal to the difference in energy between the situations (a) and (b) pictured in Fig.4.10. Situation (a) is of higher energy than (b), but corresponds to a much more asymmetric repartition of charge. It will therefore lead to a stronger polarization of the neighboring bonds, so that its energy will be more strongly lowered. A rough estimate of this effect can be obtained as follows: U is in fact equal to the energy of the dipolar distribution pictured by situation (c). The effective value of U will thus correspond to the energy of the dipole in contact with a medium of dielectric constant ε. In a bulk material, classical reasoning would lead to a reduction of this energy by ε. Here, the situation is intermediate between bulk and vacuum, so that an order of magnitude estimate of this reduction could be $(1+\varepsilon)/2$ (i.e., 3.3 for C and 6.3 for Si). The effective values for U will then become U (diamond) $\simeq$ 3.3 eV and U (silicon) $\simeq$ 1 eV (corresponding to nonreduced values of 11 and 6.5 eV, respectively). That these values provide a correct order of magnitude can also be demonstrated by another method [4.6].

Let us finally discuss the results of the defect molecule models with these effective values for the parameters. For diamond ($\Delta \simeq$ 4.3 eV, U $\simeq$ 3.3 eV) the ratio Δ/U takes the value 1.29 while, for silicon ($\Delta \simeq$ 1.2 eV, U $\simeq$ 1 eV) it becomes equal to 1.1. In both cases, we are at the limit of the region in Fig.4.9 where one-electron treatments become valid. The importance of correlation effects has been studied in [4.39] concentrating on the singlet-triplet splitting within the $v^2 t^2$ configuration of V^0. This was done by the self-consistent-field X_α-scattered wave method. The main result is that the triplet $^3T_1$ lies lower than the average energy of the zero spin states taken as $1/9$ $(3 \times {}^3T_1 + 3 {}^1T_2 + 2 {}^1E + {}^1A_1)$ and that this energy difference decreases when the size of the cluster increases. From Table 4.3, we readily find that, in the model the $^3T_1$ state is lower by $U/4$ in agreement with MESSMER and WATKINS result [4.39]. Their value for this splitting leads to U of the order of 1.2 to 2.4 eV (for a 16 atom cluster $C_4 H_{12}$ or C_{12}), which seems rather small compared to 3.3 eV for U in diamond.

5. Vibrational Properties and Entropy

Point defects have an influence on the vibrational properties of solids. They can induce localized states of vibration exactly as for the electronic spectrum. An important category of such defects are substitutional impurities with a mass lighter than the host atoms. They are called mass defects and the frequency of their localized modes gives information on the coupling between the impurity and its neighbors. Experimentally, the changes in the phonon spectrum are seen, for instance, in infrared absorption. In this chapter we discuss these properties on the basis of simple models. We also apply these models in order to estimate the vibrational contribution to the entropy, which is important for the knowledge of the thermodynamic properties of defects.

We first give a simple and brief overview of the theory of atomic vibrations, or phonons [5.1-4]. After having recalled the Born-Oppenheimer and the harmonic approximations, we solve the vibrational Hamiltonian for a linear chain. We extend this treatment to three dimensional systems and make a specific application to the case of covalently bonded solids. In regard to defects, we mainly discuss the "mass defect" with the help of very simple mathematical methods. Following this, we present the different experimental techniques and discuss some of their results in the light of the theoretical considerations.

A second part is devoted to a discussion of the vibrational entropy due to elementary defects. This is a difficult theoretical problem because it necessitates the knowledge of the frequencies of vibration of the atoms about their equilibrium positions. For a defect this means that we must know the relaxation and distortion and their effect on phonons. Consequently, few authors have studied the effect of a vacancy on the vibration spectrum in cubic lattices. However, the general feeling is that the creation of a vacancy only leads to a small increase in entropy, of the order of k. As we shall show, this increase is due to the breaking of bonds, which leads to a decrease of the vibration frequencies. This entropy change can be large when atomic displacements are induced around the defect. In such case, the force constants can be substantially modified, leading to a strong change in entropy.

5.1 Vibrational Modes

We discuss here the solutions of the Schrödinger equation for atomic displacements. We introduce the dynamical matrix and derive solutions in simple cases. Among them we treat the linear chain with one and two atoms per unit cell as well as the covalent lattice.

5.1.1 The Dynamical Matrix

We make use of the Born-Oppenheimer approximation which separates the motion of the electrons and the nuclei. We thus assume that we know the electronic energy $E_{el}(\{R\})$ for a given set of positions R of the nuclei. This energy plays the role of a potential energy for the motion of the nuclei. We have to solve the Schrödinger equation

$$\left[-\frac{\hbar^2}{2} \sum_i \frac{\Delta_i}{M_i} + E_{el}(\{R\})\right]\chi = E\chi \quad , \tag{5.1}$$

where M_i and Δ_i are respectively the mass and Laplacian operator corresponding to atom i; E and χ are respectively the total energy and vibration eigenfunction.

The general resolution of this equation is too complex. For small atomic displacements $u_{i\alpha}$ (the αth component of the displacement of atom i) about the equilibrium configuration $\{R_0\}$, we can expand $E_{el}(R)$ to second order. The equations becomes

$$\left(-\frac{\hbar^2}{2} \sum_{i,\alpha} \frac{1}{M_i} \frac{\partial^2}{\partial u_{i\alpha}^2} + \frac{1}{2} \sum_{i,j} A_{i\alpha,j\beta} u_{i\alpha} u_{j\beta}\right)\chi = \epsilon\chi \quad , \tag{5.2}$$

where ϵ is the energy relative to the minimum $E_{el}(\{R_0\})$ and $A_{i\alpha,j\beta}$ is the matrix of the second-order derivatives of $E_{el}(\{R\})$. This matrix A is the matrix of force constants. Let us now define

$$v_{i\alpha} = \sqrt{M_i}\, u_{i\alpha} \tag{5.3}$$

and write the equation in terms of the $v_{i\alpha}$. We obtain

$$\left(-\frac{\hbar^2}{2}\sum_{i,\alpha}\frac{\partial^2}{\partial v_{i\alpha}^2} + \frac{1}{2}\sum_{i\alpha,j\beta}D'_{i\alpha,j\beta}v_{i\alpha}v_{j\beta}\right)\chi = \epsilon\chi \quad , \tag{5.4}$$

where D', given by

$$D'_{i\alpha,j\beta} = \frac{A_{i\alpha,j\beta}}{\sqrt{M_iM_j}} \quad , \tag{5.5}$$

is one possible definition of the dynamical matrix. This matrix is symmetrical and can be diagonalized. If λ and $v(\lambda)$ are respectively its eigenvalues and eigenfunctions, we have

$$\left\{\sum_\lambda\left[-\frac{\hbar^2}{2}\frac{\partial^2}{\partial v(\lambda)^2} + \frac{1}{2}\lambda v(\lambda)^2\right]\right\}\chi = \epsilon\chi \quad . \tag{5.6}$$

This is the Schrödinger equation of a set of independent harmonic oscillators, whose vibration frequencies are given by $\lambda = \omega^2$. The whole problem then consists in the diagonalization of the dynamical matrix. In vector notation ($|v\rangle$ being a vector of components $v_{i\alpha}$ in an orthonormal basis set $|v_{i\alpha}\rangle$, and $\underset{\approx}{D}'$ being considered as an operator in this vectorial space),

$$(\underset{\approx}{D}' - \omega^2\underset{\approx}{I})|v\rangle = 0 \quad , \tag{5.7}$$

where $\underset{\approx}{I}$ is the unit operator.

If we come back to the original force constant matrix A, the eigenvalue equation can be written:

$$(\underset{\approx}{A} - \omega^2\underset{\approx}{M})|u\rangle = 0 \quad , \tag{5.8}$$

with $\underset{\approx}{M}$ corresponding to a diagonal matrix of elements M_i. A third, and sometimes quite useful form (especially for the mass defect) for the same eigenvalue equation, is the following:

$$(\underset{\approx}{D} - \omega^2\underset{\approx}{I})|u\rangle = 0 \quad . \tag{5.9}$$

Here $\underset{\approx}{D}$, equal to $\underset{\approx}{M}^{-1}\underset{\approx}{A}$, is another definition of the dynamical matrix. An important property of the force constants matrix $\underset{\approx}{A}$ results from translational invariance. To show this, let us write the expression of $F_{i\alpha}$, the αth component of the force on atom i. It is given by

$$F_{i\alpha} = -\sum_{j\beta}A_{i\alpha,j\beta}u_{j\beta} \quad . \tag{5.10}$$

This force component is left unchanged by any translation T of the whole system. The choice of a translation such that $u_{j\beta}$ is equal to T, for any j and fixed β, leads to

$$\sum_j A_{i\alpha,j\beta} = 0 \quad . \tag{5.11}$$

This relation is useful to reduce the number of independent parameters.

5.1.2 The Linear Chain

As an example, we treat the linear chain with one atom per unit cell, the force constants being restricted to nearest neighbors (see Fig.5.1). We consider displacements u_i along the chain, and write any eigenvector $|u\rangle$ under the form

$$|u\rangle = \sum_i u_i |u_i\rangle \quad . \tag{5.12}$$

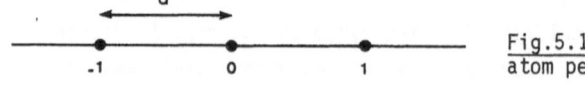

Fig.5.1. The linear chain with one atom per unit cell

The translational periodicity allows us to use the Bloch theorem and classify the eigenstates by their wave vector k, so that

$$|u(k)\rangle = \frac{1}{\sqrt{N}} \sum_j e^{ikja} |u_j\rangle \quad , \tag{5.13}$$

N being the number of atoms. Since all atoms have an identical mass M, we can write the secular equation as

$$\frac{1}{M} [A_0 - \omega^2(k)] |u(k)\rangle = 0 \quad . \tag{5.14}$$

The projection of this equation onto the basis vector $|u_0\rangle$ leads to

$$\omega^2(k) = \frac{1}{M} \sum_j e^{ikja} A_{0j} \quad . \tag{5.15}$$

The A_{0j} terms decrease when j increases. The simplest model consists then to consider only the nearest neighbor interactions, in which case we can write by symmetry

$$A_{01} = A_{0-1} = -C \quad . \tag{5.16}$$

Using the property of translational invariance given by (5.11) we obtain

$$A_{00} = -(A_{01} + A_{0-1}) = 2C \quad . \tag{5.17}$$

The frequencies $\omega(k)$ can thus be expressed in terms of a single force constant C, with

$$\omega(k) = 2 \sqrt{\frac{C}{M}} \left| \sin \frac{ka}{2} \right| \quad . \tag{5.18}$$

The corresponding dispersion relation $\omega(k)$ is drawn in Fig.5.2a. The slope at $k = 0$ gives the velocity of acoustic waves. The maximum frequency ω_M is given by $2\sqrt{C/M}$. The density of states $n(\omega)$ corresponding to (5.18) is determined by

$$n(\omega)d\omega = \nu(k)dk = \frac{L}{\pi} dk \quad . \tag{5.19}$$

Therefore,

$$n(\omega) = \frac{L}{\pi} \frac{dk}{d\omega} = 2 \frac{N}{\pi} \frac{1}{\sqrt{\omega_M^2 - \omega^2}} \quad , \tag{5.20}$$

where L is the length of the chain, and N the number of atoms.

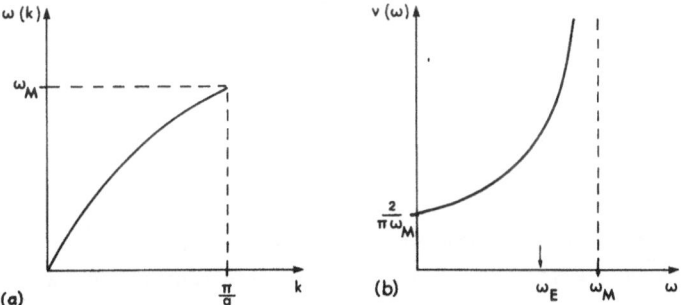

Fig.5.2a,b. Phonon dispersion curve for a linear chain with one atom per unit cell (a) and corresponding density of states (b)

The density of states per atom $\nu(\omega)$, drawn in Fig.5.2b, is then equal to

$$\nu(\omega) = \frac{2}{\pi} \frac{1}{\sqrt{\omega_M^2 - \omega^2}} \tag{5.21}$$

On this figure we indicate the Einstein frequency ω_E. It corresponds to an approximation where the nondiagonal terms of A are not considered. The eigenvalues of A are thus equal to its diagonal terms. They are all degenerate, leading to a common eigenfrequency,

$$\omega_E = \sqrt{\frac{2C}{M}} \quad , \tag{5.22}$$

that is

$$\omega_E = \frac{\omega_M}{\sqrt{2}} \quad . \tag{5.23}$$

This simplified model will turn out to be very useful.

5.1.3 The Linear Chain with Two Atoms Per Unit Cell

This example is useful because it shows the splitting of the vibration frequencies, into acoustical and optical modes. Let us consider the chain of Fig.5.3 with two atoms A and B (of mass M_A and M_B) per unit cell. The length of the unit cell is now 2a. We adopt the same model of nearest-neighbors force constants, as in Sect.5.1.2, given by (5.16,17). In order to use Bloch's the-

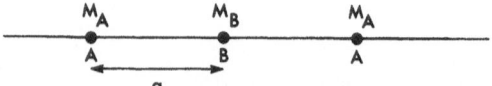

Fig.5.3. The linear chain with two atoms per unit cell

orem we define the Bloch sums:

$$|u_A(k)> = \frac{1}{\sqrt{N}} \sum_j e^{2ikja} |u_A(j)>$$

$$|u_B(k)> = \frac{1}{\sqrt{N}} \sum_j e^{2ikja} |u_B(j)> \quad , \tag{5.24}$$

where $|u_A(j)$ and $|u_B(j)>$ are the basis vectors which correspond respectively to the atoms A and B of the jth unit cell. We write any eigenstate of wave number k as a linear combination of $u_A(k)$ and $u_B(k)$:

$$|u(k)> = C_A(k)|u_A(k)> + C_B(k)|u_B(k)> \quad . \tag{5.25}$$

We now project the general eigenvalue equation (5.8) on the basis vectors $|u_A(0)>$ and $|u_B(0)>$. This leads to the set of equations

$$<u_A(0)|(A - M_A\omega^2)|u_A(k)>C_A(k) + <u_A(0)|A|u_B(k)>C_B(k) = 0$$

$$<u_B(0)|A|u_A(k)> C_A(k) + <u_B(0)|(A - M_B\omega^2)|u_B(k)> C_B(k) = 0 \quad , \tag{5.26}$$

which, in the simple force constant model adopted here, reduces to

$$(2C - M_A\omega^2)C_A(k) - C(1 + e^{-2ika})C_B(k) = 0$$

$$- C(1 + e^{2ika})C_A(k) + (2C - M_B\omega^2)C_B(k) = 0 \quad . \tag{5.27}$$

The determinant of the coefficients must vanish. This leads to the condition

$$\omega^2(k) = C\left[\frac{1}{M_A} + \frac{1}{M_B} \pm \sqrt{\left(\frac{1}{M_A} + \frac{1}{M_B}\right)^2 - \frac{4\sin^2 ka}{M_A M_B}}\right] \quad . \tag{5.28}$$

The resulting curve, $\omega(k)$ versus k, is drawn in Fig.5.4. It shows, that there are now two branches in the dispersion relation (they correspond to the case $M_A > M_B$).

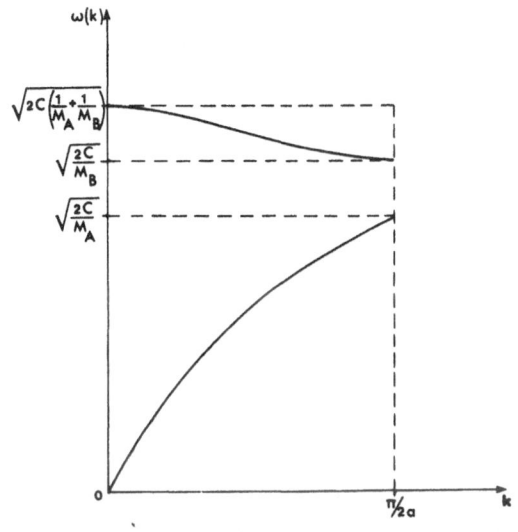

Fig.5.4. Dispersion curve for a linear chain with two atoms per unit cell

Let us now look at the atomic displacements. These are the components of $|u(k)\rangle$ along the basis vectors $|u_A(j)\rangle$ and $|u_B(j)\rangle$ and are thus equal to $C_A(k)\exp(2ikja)$ and $C_B(k)\exp(2ikja)$. Their ratio, in the long wavelength limit ($k \to 0$), is given (using (5.27)) by

$$\frac{C_B(0)}{C_A(0)} = 1 - \frac{M_A}{2C}\omega^2 \quad . \tag{5.29}$$

For the lower branch, $\omega^2 \to 0$, it tends to unity. It corresponds to the propagation of acoustic waves. For the upper branch, ω^2 tends towards $2C(1/M_A + 1/M_B)$ and the ratio tends towards $-M_A/M_B$. The two kinds of atoms vibrate in opposite directions. This corresponds to vibrating dipoles which can couple with an electromagnetic field, resulting in infrared adsorption (this is discussed in Sect.5.3). For this reason, the upper branch is called the optical branch. In the Einstein model, which only retains the diagonal part of A (Sect.5.1.2), there are two Einstein frequencies,

$$\omega_{EA} = \sqrt{\frac{2C}{M_A}} \quad \text{and} \quad \omega_{EB} = \sqrt{\frac{2C}{M_B}} \quad . \tag{5.30}$$

These two frequencies are just equal to the limits of the optical and acoustical branches.

5.1.4 The Covalent Solid

There have been many calculations of the phonon dispersion curves of the Group IV covalent systems, as well as the III-V and II-VI compounds. They can be classified in three categories, depending upon the method used to build the phonon hamiltonian: a) the valence force field model (originally applied to molecules), b) the shell model and c) the dielectric approach. Here we shall develop a simple version of the valence force-field model which will serve as an illustration of a three-dimensional system with two atoms per unit cell. It will also be useful for our discussions of entropies, in the second part of this chapter.

In Group IV tetrahedral covalent systems, such as C, Si, Ge, the valence force-field method corresponds to the following expression of the energy, developed to second order in the displacements:

$$E_{el}(\{\underset{\sim}{R}\}) = E_{el}(\{\underset{\sim}{R_0}\}) + \frac{1}{2} k_{rr} \sum_{ij} (dr_{ij})^2 + \frac{1}{2} k_\theta \sum_\alpha (d\theta_\alpha)^2$$

$$+ \frac{1}{2} k_{r\theta} \sum_{i\alpha} (dr_{ij} d\theta_\alpha) + \dots \tag{5.31}$$

The dr_{ij} are the changes in bond length and the $d\theta_\alpha$ the changes in bond angles. This expansion can include as many cross terms $dr_i d\theta_\alpha$ or $dr_{ij} dr_{k\ell}$ as is needed. Here we only consider the effects of the first two terms, i.e., bond stretching and bond bending, which already give a good account of the phonon dispersion curves $\omega(k)$ except for some fine details [5.5]. The extension of this model to III-V and II-VI compounds requires the addition of long-range electrostatic effects, which lead to the "Lyddane-Sachs-Teller" splitting. This will be discussed at the end of this section.

To begin, we consider the first term $0.5 \, k_{rr} \sum_{ij} (dr_{ij})^2$ alone. It is trivial to demonstrate that the intraatomic elements of the force constants matrix A are

$$A_{i\alpha,i\beta} = k_{rr} \sum_j x_{ij}^\alpha x_{ij}^\beta \quad , \tag{5.32}$$

where j are the nearest neighbors of i; X_{ij}^α is the αth component of the unit vector $\hat{n}_{ij}$ along the bond i - j. In the same manner, the nearest neighbors' interatomic term is

$$A_{i\alpha,j\beta} = - k_{rr} X_{ij}^\alpha X_{ij}^\beta \quad . \tag{5.33}$$

The eigenvalue equation can thus be written

$$M\omega^2 u_{i\alpha} = k_{rr} \sum_{j\beta} X_{ij}^\alpha X_{ij}^\beta (u_{i\beta} - u_{j\beta}) \quad , \tag{5.34}$$

that is

$$M\omega^2 u_{i\alpha} = k_{rr} \sum_j X_{ij}^\alpha dr_{ij} \quad . \tag{5.35}$$

Similarly, if k is one of the nearest neighbors of i, we have

$$M\omega^2 u_{k\alpha} = k_{rr} \sum_\ell X_{k\ell}^\alpha dr_{k\ell} \quad . \tag{5.36}$$

We take the difference of these two equations, multiply by X_{ik}^α and sum over α. We obtain

$$\frac{M\omega^2}{k_{rr}} dr_{ik} = \sum_j (\hat{n}_{ik}\hat{n}_{ij}) dr_{ij} - \sum_\ell (\hat{n}_{ik}\hat{n}_{k\ell}) dr_{k\ell} \quad . \tag{5.37}$$

The scalar products, in the perfect diamond lattice, are 1 if j = k, -1/3 if j ≠ k, -1 if ℓ = k, +1/3 if ℓ ≠ k. Thus,

$$\left(\frac{M\omega^2}{k_{rr}} - \frac{8}{3}\right) dr_{ik} = -\frac{1}{3}\left(\sum_j dr_{ij} + \sum_\ell dr_{k\ell}\right) \quad . \tag{5.38}$$

Let us now label S_i the sum $\sum_j dr_{ij}$. Summing (5.38) over k gives

$$\left(\frac{M\omega^2}{k_{rr}} - \frac{8}{3}\right) S_i = -\frac{4}{3} S_i - \frac{1}{3}\sum_k S_k \quad , \tag{5.39}$$

or equivalently,

$$3\left(\frac{4}{3} - \frac{M\omega^2}{k_{rr}}\right) S_i = \sum_k S_k \quad . \tag{5.40}$$

This relation has the form of a one-band problem,

$$\epsilon S_i = \sum_k S_k \quad , \tag{5.41}$$

exactly as was the case in Chap.3 for the tight-binding electronic Hamiltonian. The same considerations apply, leading to

164

$$\omega^2 = \frac{k_{rr}}{M}\left(\frac{4-\varepsilon}{3}\right) \quad , \tag{5.42}$$

where the dispersion relation for ε depends on the lattice (for the tetrahedral coordination ε lies in the interval [-4, +4]). There are additional solutions corresponding to $S_i = 0$. (One of them occurs when $M\omega^2/k_{rr} = 8/3$.) Finally, there is the possibility $\omega^2 = 0$, which requires that all dr_{ik} be zero, i.e., corresponds to transverse waves. The density of states $n(\omega^2)$ is plotted in Fig.5.5. The two flat bands contain 1 state per atom, the broad band r_{00}, leading to a total of three states per atom as required. These results correspond to those derived in [5.6].

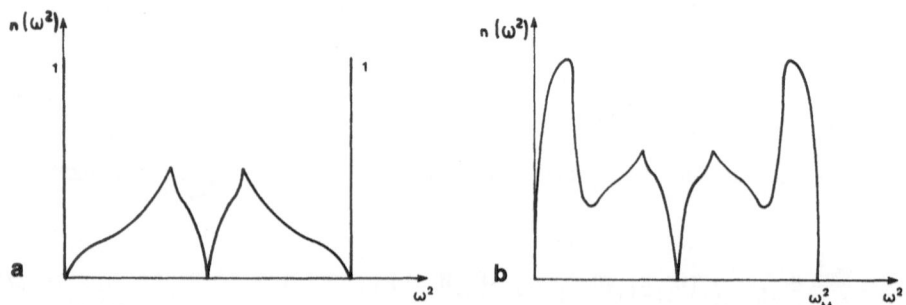

Fig.5.5a,b. Phonon density of states in a covalent material when only bond stretching terms are considered (a) and when bond bending terms are added (b)

Corrections to this extremely simple description give the effect described in Fig.5.5b, which corresponds to the inclusion of k_θ (bond bending terms). Clearly, there is not much fundamental change, except the broadening of the two flat bands.

The main difference which exists between the vibration spectra of the Group IV elements and of the III-V or the II-VI compounds is given schematical in Fig.5.6a,b. The dispersion relation $\omega(\underline{k})$ is given for $\underline{k}$ along the high symmetry axes, the (100) and (111) directions. Let us consider first Fig.5.6a corresponding to the Group IV elements. We have labelled ω_{LO} and ω_{LA} the optic and acoustic branches, corresponding to longitudinal waves, for which the atomic displacements are along the wave vector; ω_{LA} and ω_{TA}, on the other hand correspond to transverse waves, for which the atomic displacements are perpendicular to the wave vector. The fact that pure longitudinal and pure transverse solutions are obtained is a consequence of the high symmetry.

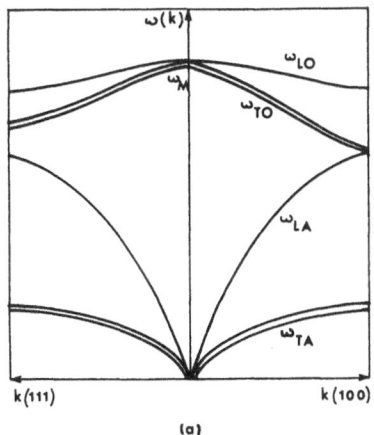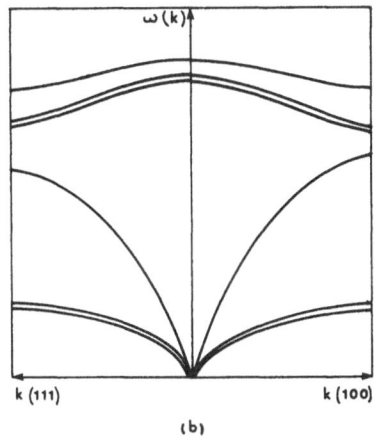

Fig.5.6a,b. Dispersion curve for a group IV material (a) and for a III-V or II-VI compound (b)

For a zinc-blende compound there is, as shown in Fig.5.6b, the same number of branches, the number of atoms in the unit cell being unchanged. However, at the limit of the Brillouin Zone, the degeneracies are lifted due to the mass difference between the two atoms. Another striking difference comes from the splitting of ω_{LO} and ω_{TO} at $\underline{k} = 0$, the "Lyddane-Sachs-Teller splitting". This splitting results from the fact that, in ionic systems, longitudinal displacements at $\underline{k} = 0$ induce a macroscopic electric field, which is not the case for transverse displacements [5.7].

5.2 Localized Modes Due to Defects

We now develop different models corresponding to point defects. We will first discuss the mass defect for the linear chain, next the mass defect for three-dimensional systems by expanding the Green's functions and, finally, develop an application to the vacancy.

5.2.1 The Mass Defect in the Monoatomic Linear Chain

Consider a substitutional impurity which does not alter the local force constants; this perturbation corresponds to a change in the mass of the corresponding atom. Although this problem can be handled exactly with.the Green's functions method derived in Chap.3, it is interesting to first solve a simpler case, i.e., the linear chain. In this case, we can write (5.8) projected onto the different basis vectors $|u_0>, |u_1> \ldots$, corresponding to the different atoms. This leads to the following set of equations:

$$(M\omega^2 - 2C)u_0 = - C(u_1 + u_{-1})$$

$$(M_0\omega^2 - 2C)u_1 = - C(u_0 + u_2)$$

$$(M_0\omega^2 - 2C)u_{-1} = -C(u_0 + u_{-2})$$

$$(M_0\omega^2 - 2C)u_2 = -C(u_1 + u_3)$$

$$(M_0\omega^2 - 2C)u_{-2} = -C(u_{-1} + u_{-3}) \quad , \qquad (5.43)$$

where M is now the impurity mass and M_0 the unperturbed mass. It can be seen by inspection that a solution of this set of equations can be written in the form,

$$u_i = AK^{|i|} \quad , \qquad (5.44)$$

with the two conditions,

$$M_0\omega^2 - 2C = -C\left(K + \frac{1}{K}\right) \qquad (5.45)$$

and

$$M\omega^2 - 2C = -2KC \quad , \qquad (5.46)$$

which provide for ω^2 the solution,

$$\omega = \frac{\omega_M}{\sqrt{1-\varepsilon^2}} \quad , \qquad (5.47)$$

where

$$\varepsilon = 1 - \frac{M}{M_0} \qquad (5.48)$$

is a quantity called the mass defect. Clearly, when M is smaller than M_0, there exists a solution for ω which lies above the maximum frequency ω_M of the perfect crystal. The nature of this new eigenstate can be analyzed by solving (5.45,46) for K using ω^2 given by (5.47). This gives

$$K = - \frac{1-\varepsilon}{1+\varepsilon} = - \frac{M/M_0}{2-M/M_0} \quad . \qquad (5.49)$$

For $\varepsilon<1$, $|K|$ is always smaller than unity, allowing us to rewrite (5.44) for the displacements u_i of atom i in the form

$$u_i = (-1)^{|i|} \exp(+|i| \log|K|) \quad . \tag{5.50}$$

Since $\log|K|$ is negative, the atomic displacements decay exponentially away from the defect, this decay becoming faster as M/M_0 decreases. This is called a localized mode of vibration.

A substitutional impurity will more or less alter the force constants in its neighborhood. To investigate their effect on the localized mode frequency, we assume that only the coupling of the impurity with its nearest neighbors is altered, C being changed into $C' = C + \Delta C$. Only the first three equations of (5.43) are changed and become

$$(M\omega^2 - 2C')u_0 = -C'(u_1 + u_{-1})$$

$$(M_0\omega^2 - C - C')u_1 = -(C'u_0 + Cu_2)$$

$$(M_0\omega^2 - C - C')u_{-1} = -(C'u_0 + Cu_{-2}) \quad . \tag{5.51}$$

We again search for a solution of the form given by (5.44) for the u_i with $|i| \geq 1$. This means that (5.45) must be added to this set of equations (5.51). Because the solution is symmetrical with respect to the impurity site, we have to solve

$$(M\omega^2 - 2C')u_0 = -2C'AK$$

$$(M_0\omega^2 - C - C')AK = -(C'u_0 + CAK^2)$$

$$(M_0\omega^2 - 2C) = -C(K + \frac{1}{K}) \quad . \tag{5.52}$$

By eliminating u_0 from the first two equations, we end up with

$$\left(M_0\omega^2 - C - C' - \frac{2(C')^2}{M\omega^2 - 2C'}\right) = -CK$$

$$M_0\omega^2 - 2C = -C(K + \frac{1}{K}) \quad . \tag{5.53}$$

Since the resolution of this set of equations is lengthy, we will only outline the method. From the second equation, we express K under the form

$$K = 1 - \frac{M_0\omega^2}{2C} \pm \sqrt{\left(\frac{M_0\omega^2}{2C}\right)^2 - \frac{M_0\omega^2}{C}} \tag{5.54}$$

and inject this value into the first one. An intermediate step is given by

$$\left(\frac{M_0\omega^2}{2} - C' - \frac{2C'^2}{M\omega^2 - 2C'}\right) = \pm\sqrt{\left(\frac{M_0\omega^2}{2}\right)^2 - M_0\omega^2 C} \quad . \tag{5.55}$$

By taking the square of the two members and after some manipulations, an equation of second order in ω^2 is obtained. Its coefficients are expressed in terms of ε, the mass defect, and $x = \Delta C/C$, the change in the force constant. The final result can be cast under the form

$$\omega^2 = \frac{\omega_M^2}{1-\varepsilon^2} \cdot \frac{1+x}{\left(\frac{1}{2} + \frac{x}{2}\frac{(1-\varepsilon)^2}{1-\varepsilon^2}\right) + \sqrt{\left(\frac{1}{2} + \frac{x}{2}\frac{(1-\varepsilon)^2}{1-\varepsilon^2}\right)^2 - 4x\left(\frac{1-\varepsilon}{1-\varepsilon^2}\right)^2}} \quad . \tag{5.56}$$

Let us consider the case of an increase in force constants, i.e., $x > 0$. The term $(1 - \varepsilon)^2/(1 - \varepsilon^2)$ is smaller than unity, the denominator is thus smaller than $1 + x$, and the frequency of the local mode is higher than its previous value. A symmetrical argument is valid for $x < 0$. We shall see later that this result is fairly easy to understand.

5.2.2 The Mass Defect in the General Case

We shall now discuss the general method of obtaining the local mode frequencies for a mass defect [5.8]. This is again the Green's function method which was described in Chap.3. All the theory presented there can be applied here simply by changing H into the dynamical matrix D (or D'), defined previously, and E into ω^2. The resolvent operator for the perfect crystal is defined by

$$\mathbf{G}_0 = (\omega^2 + i\eta - \mathbf{D}_0)^{-1} \quad , \tag{5.57}$$

$\mathbf{D}_0$ being the dynamical matrix, which we take equal to

$$\mathbf{D}_0 = \mathbf{M}_0^{-1}\mathbf{A}_0 \quad , \tag{5.58}$$

the subscript zero referring to unperturbed operators. The local modes are again solutions of the equation

$$\det|\mathbf{I} - \mathbf{G}_0 \mathbf{V}| = 0 \quad , \tag{5.59}$$

where $\mathbf{V}$ is the perturbation introduced by the mass defect. The most simple method consists of expressing $\mathbf{V}$ as the difference between $\mathbf{D}$ and $\mathbf{D}_0$, i.e., $\mathbf{M}^{-1}\mathbf{A}$ and $\mathbf{M}_0^{-1}\mathbf{A}_0$. If we assume a pure mass defect ($\mathbf{A} = \mathbf{A}_0$), $\mathbf{V}$ is the following matrix:

$$\underset{\sim}{V} = (\underset{\sim}{M}^{-1} - \underset{\sim}{M_0}^{-1})\underset{\sim}{A_0} \quad , \tag{5.60}$$

whose nonvanishing matrix elements are given by

$$V_{0\alpha,i\beta} = \left(\frac{1}{M} - \frac{1}{M_0}\right)\left(A_0\right)_{0\alpha,i\beta} \tag{5.61}$$

The subscripts 0 and i are atomic indices and α and β refer to the components of the displacements. Even though the mass defect occurs on the site 0, the perturbation has the same extension as A_0 itself. The determinant (5.59) is therefore lengthy to calculate. A much more direct method can be used, starting from (5.8), written for the perturbed system:

$$\left(\underset{\sim}{A_0} - \underset{\sim}{M}\omega^2\right)|u\rangle = 0 \quad . \tag{5.62}$$

In this equation, we add and subtract $\underset{\sim}{M_0}\omega^2$, multiply by $\underset{\sim}{M_0}^{-1}$, and obtain

$$\left[\underset{\sim}{D_0} - \left(\underset{\sim}{M_0}^{-1}\underset{\sim}{M} - \underset{\sim}{I}\right)\omega^2 - \omega^2\underset{\sim}{I}\right]|u\rangle = 0 \quad . \tag{5.63}$$

With this formula, $\underset{\sim}{V}$ becomes

$$\underset{\sim}{V} = -\left(\underset{\sim}{M_0}^{-1}\underset{\sim}{M} - \underset{\sim}{I}\right)\omega^2 \quad , \tag{5.64}$$

whose only nonvanishing elements are

$$V_{0\alpha,0\alpha} = \left(1 - \frac{M}{M_0}\right)\omega^2 \quad , \tag{5.65}$$

that is,

$$V_{0\alpha,0\alpha} = \varepsilon\omega^2 \quad . \tag{5.66}$$

This form of $\underset{\sim}{V}$ leads to an easier calculation of the determinant. Let us first apply it to the monoatomic linear chain, with displacements along the chain, so that we can drop the index α and β. The problem is completely equivalent to the Slater-Koster one (see Chap.3) and (5.59) reduces to

$$1 - \varepsilon\omega^2\left(G_0\right)_{00} = 0 \quad . \tag{5.67}$$

The Green's function $(G_0)_{00}$ can be calculated in the following way

$$\left(G_0\right)_{00} = \langle u_0|(\omega^2 + i\eta - \underset{\sim}{D_0})^{-1}|u_0\rangle \quad , \tag{5.68}$$

that is

$$\left(G_0\right)_{00} = \sum_k \frac{<u_0|u(k)> <u(k)|u_0>}{\omega^2+i\eta-\omega^2(k)} \quad , \tag{5.69}$$

and finally

$$\left(G_0\right)_{00} = \frac{1}{N} \sum_k \frac{1}{\omega^2+i\eta-\omega^2(k)} \quad . \tag{5.70}$$

We wish to search for localized states such that $\omega > \omega_M$. In this range $(G_0)_{00}$ is real and given by

$$\left(G_0\right)_{00} = \frac{1}{N} \sum_k \frac{1}{\omega^2-\omega^2(k)} \quad , \tag{5.71}$$

that is

$$\left(G_0\right)_{00} = \frac{a}{2\pi} \int_{-\pi/a}^{+\pi/a} \frac{dk}{\omega^2-\omega^2(k)} \quad , \tag{5.72}$$

and

$$\left(G_0\right)_{00} = \frac{1}{|\omega|} \frac{1}{\sqrt{\omega^2-\omega_M^2}} \quad . \tag{5.73}$$

The combination of (5.73) with (5.67) leads to exactly the same expression obtained by the direct method.

The mass defect in three-dimensional systems can be treated simply if there is enough symmetry to write

$$\left(G_0\right)_{0\alpha,0\beta} = \left(G_0\right)_{0\alpha,0\alpha}\delta_{\alpha\beta} \quad . \tag{5.74}$$

This occurs, for instance, in cubic systems where the three $\left(G_0\right)_{0\alpha,0\alpha}$ (with $\alpha = x,y,$ or z) are identical. In this case, the determinant (5.59) factorizes and we obtain

$$1 - \omega^2\left(G_0\right)_{0\alpha,0\alpha} = 0 \quad . \tag{5.75}$$

If a localized mode, given by (5.75), exists in a cubic system, then it has threefold degeneracy. In this case, however, the calculation of the Green's function must be made numerically. It is thus advantageous to derive techniques which allow this calculation to be performed by successive approximations. They are based on moments expansions and give analytical results with sufficient accuracy to allow simple physical discussions.

5.2.3 Expansion of the Green's Function

The method consists in expanding the Green's functions in powers of ω^{-2} for $\omega > \omega_M$, the maximum frequency of the phonon spectrum. For this, we write the dynamical matrix $\underset{\approx}{D}$ as the sum of its diagonal part $\underset{\approx}{d}$ and its nondiagonal part $\underset{\approx}{R}$ and we define the resolvent $\underset{\approx}{g}$ of the diagonal part as

$$\underset{\approx}{g} = \left(\omega^2 - \underset{\approx}{d}\right)^{-1} \quad . \tag{5.76}$$

Using Dyson's equation to relate $\underset{\approx}{G}$, the resolvent of $\underset{\approx}{D}$, to $\underset{\approx}{g}$

$$\underset{\approx}{G} = \underset{\approx}{g} + \underset{\approx}{g}\,\underset{\approx}{R}\,\underset{\approx}{G} \quad , \tag{5.77}$$

equivalent to

$$\underset{\approx}{G} = \left(\underset{\approx}{1} - \underset{\approx}{g}\,\underset{\approx}{R}\right)^{-1}\underset{\approx}{g} \quad , \tag{5.78}$$

we obtain

$$\underset{\approx}{G} = \sum_{n=0} \underset{\approx}{g}\left(\underset{\approx}{R}\,\underset{\approx}{g}\right)^n \quad . \tag{5.79}$$

The first term in this expansion corresponds to the Einstein approximation in which each atom is considered as an isolated oscillator. This can be seen by taking a diagonal matrix element of $\underset{\approx}{G}$:

$$<u_{i\alpha}|\underset{\approx}{G}|u_{i\alpha}> = \frac{1}{\omega^2 - d_{i\alpha,i\alpha}} \quad . \tag{5.80}$$

The eigenfrequencies correspond to the poles of the Green's function, wich are given by

$$\omega^2 = d_{i\alpha,i\alpha} \quad . \tag{5.81}$$

This expression, which was introduced in Sect. 5.1.2, has the advantage of defining the Einstein approximation as the first step in a power-series expansion.

Let us now consider the mass defect for the linear chain in this Einstein model. The impurity atom will vibrate at the frequency

$$\omega = \sqrt{\frac{2C}{M}} \quad , \tag{5.82}$$

that is

$$\omega = \frac{\omega_M}{\sqrt{2(1-\epsilon)}} \quad . \tag{5.83}$$

This last expression is equivalent to the exact one (5.47) when ε tends towards unity, i.e., $M = 0$, the limit of very high frequency for the local modes. When $\varepsilon \to 0$, it gives $\omega_M/\sqrt{2}$ instead of ω_M, which corresponds to an error by about 30%. We can go beyond the Einstein approximation by including higher-order terms in the expansion. Let us then calculate the matrix element $G_{00} = \langle u_0 | \underset{\approx}{G} | u_0 \rangle$ of the Green's function on the impurity site. The application of (5.79) leads to

$$G_{00} = g_{00} + \sum_i g_{00} R_{0i} g_{ii} R_{i0} g_{00} + \cdots \quad . \tag{5.84}$$

Since our definition of $\underset{\approx}{D}$ is $\underset{\approx}{M}^{-1} \underset{\approx}{A}$ (see Sect.5.1.1), we have the following relations:

$$g_{00} = \frac{1}{\omega^2 - \frac{2C}{M}} \quad , \quad g_{ii} = \frac{1}{\omega^2 - \frac{2C}{M_0}} \quad \text{for} \quad i \neq 0 \quad ,$$

$$R_{0i} = -\frac{C}{M} \quad , \quad \text{and} \tag{5.85}$$

$$R_{i0} = -\frac{C}{M_0} \quad .$$

The expression (5.84) therefore becomes

$$G_{00} = \frac{1}{\omega^2 - \frac{2C}{M}} \left[1 + \frac{\frac{2C^2}{MM_0}}{\left(\omega^2 - \frac{2C}{M}\right)\left(\omega^2 - \frac{2C}{M_0}\right)} + \cdots \right] \quad . \tag{5.86}$$

Equation (5.86) can be rewritten as a continuous fraction expansion to the same order of accuracy:

$$G_{00} = \left[\omega^2 - \frac{\omega_M^2}{2(1-\varepsilon)} \right]^{-1} \left[1 - \frac{\frac{\omega_M^4}{8(1-\varepsilon)}}{\left[\omega^2 - \frac{\omega_M^2}{2(1-\varepsilon)}\right]\left(\omega^2 - \frac{\omega_M^2}{2}\right)} \right]^{-1} \quad . \tag{5.87}$$

Finally,

$$G_{00} = \frac{\omega^2 - \frac{\omega_M^2}{2}}{\left(\omega^2 - \frac{\omega_M^2}{2}\right)\left[\omega^2 - \frac{\omega_M^2}{2(1-\varepsilon)}\right] - \frac{\omega_M^4}{8(1-\varepsilon)}} \quad . \tag{5.88}$$

The highest frequency pole of this Green's function can be written as

$$\frac{\omega}{\omega_M} = \left\{ \left[\frac{1}{2(1-\varepsilon)} \right] \left[\left(1 - \frac{\varepsilon}{2} \right) + \sqrt{\left(1 - \frac{\varepsilon}{2} \right)^2 - \frac{1-\varepsilon}{2}} \right] \right\}^{\frac{1}{2}} . \tag{5.89}$$

This expression has the correct limiting behavior when $\varepsilon \to 1$, which evidently is also correct for the Einstein approximation itself. In the limit $\varepsilon \to 0$, ω is equal to $0.924 \, \omega_M$ while the exact value is ω_M. The error is now reduced to 7%.

Since the Einstein model already provides a reasonable value of the localized mode (for the linear chain it becomes useful for $\varepsilon > 0.5$), it is interesting to apply it to more general situations. The localized mode frequency (if it exists) is then equal to [see (5.81)]

$$\omega_E = \sqrt{\frac{A_{0\alpha,0\alpha}}{M}} , \tag{5.90}$$

and the ratio of this frequency to the bulk Einstein frequency can be written as

$$\frac{\omega_E}{\omega_{E_0}} = \sqrt{\frac{1 + X_{\alpha,\alpha}}{1 - \varepsilon}} , \tag{5.91}$$

where $X_{\alpha,\alpha}$ is the relative change in the diagonal term of the force constant matrix and ε the mass defect. This model explains the following physical trends:

- the local mode frequency increases when the mass defect increases towards unity;
- the local mode frequency decreases when the force constants around the impurity are lower than the bulk values.

5.2.4 The Vacancy

For obvious reasons of simplicity, we only treat the vacancy in the linear chain. This simple model is pictured in Fig.5.7a, where the elements of the dynamical matrix are indicated. When the vacancy is introduced (on site 0) the problem becomes equivalent to the problem of two independent semiinfinite chains. We consider only the right part of the chain and we calculate the Green's function G_{11} by using the method detailed in Chap.3.

Dyson's equation relates G to any unperturbed g, V being the perturbation:

$$G = g + gVG . \tag{5.92}$$

We first consider Fig.5.7b to be the unperturbed problem of resolvent G_b and wish to calculate $(G_a)_{11}$ corresponding to Fig.5.7a. The perturbation is the

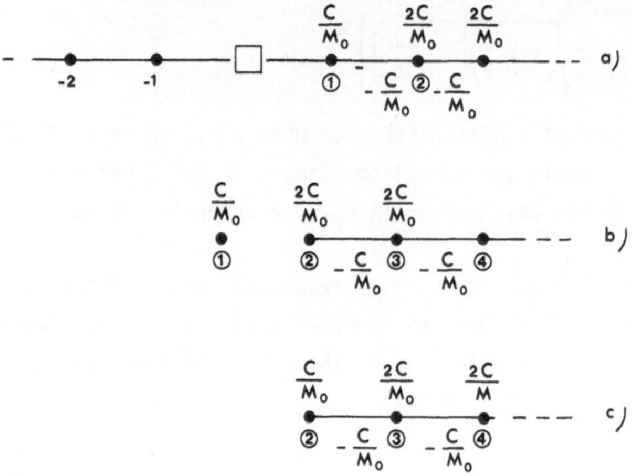

174

Fig.5.7. A linear chain containing a vacancy, equivalent to two independent linear chains. The elements of the dynamical matrix are indicated

suppression of $-C/M_0$ between sites 1 and 2. Thus

$$(G_a)_{11} = (G_b)_{11} + (G_b)_{11}V_{12}(G_a)_{21} \quad , \tag{5.93}$$

and

$$(G_a)_{21} = (G_b)_{22}V_{21}(G_a)_{11} \quad . \tag{5.94}$$

This leads to

$$(G_a)_{11} = \frac{(G_b)_{11}}{1-(G_b)_{11}V_{12}(G_b)_{22}V_{21}} \quad , \tag{5.95}$$

that is,

$$(G_a)_{11} = \frac{1}{\omega^2 - \frac{C}{M_0} - \left(\frac{C}{M_0}\right)^2 (G_b)_{22}} \quad . \tag{5.96}$$

We can do the same thing for $(G_b)_{22}$, taking Fig.5.7c as the new unperturbed problem, resulting in

$$(G_b)_{22} = (G_c)_{22} + (G_c)_{22}V_{22}(G_b)_{22} \quad , \tag{5.97}$$

that is,

$$(G_b)_{22} = \frac{(G_c)_{22}}{1 - V_{22}(G_c)_{22}} \quad , \tag{5.98}$$

and finally

$$(G_b)_{22} = \frac{(G_c)_{22}}{1 - \frac{C}{M_0}(G_c)_{22}} \quad . \tag{5.99}$$

The direct comparison between $(G_a)_{11}$ and $(G_c)_{22}$ shows that they must be equal since the physical situations are identical (see Fig.5.7). We can write, using (5.96,99),

$$(G_a)_{11} = \left[\omega^2 - \frac{C}{M_0} - \left(\frac{C}{M_0}\right)^2 \frac{(G_a)_{11}}{1 - \frac{C}{M_0}(G_a)_{11}} \right]^{-1} \quad . \tag{5.100}$$

This is a quadratic equation for $(G_a)_{11}$ whose solution is, taking into account the fact that ω_M^2 is equal to $4C/M_0$,

$$(G_a)_{11} = \frac{2}{\omega_M} \left(1 \pm \frac{1}{\omega} \sqrt{\omega^2 - \omega_M^2} \right) \quad . \tag{5.101}$$

Within the band, this gives for the density of states related to $\text{Im}\{(Ga)_{11}\}$ by (3.60),

$$(n_a)_{11} = \frac{2}{\pi} \sqrt{\frac{1}{\omega^2} - \frac{1}{\omega_M^2}} \quad . \tag{5.102}$$

The density of states is compared in Fig.5.8 to the perfect density of states $\nu(\omega^2)$ given by (5.21). This figure shows that, on the average, the vibration modes are lowered in the neighborhood of the vacancy. Also notice that there are no local modes. All this can easily be understood in the Einstein model.

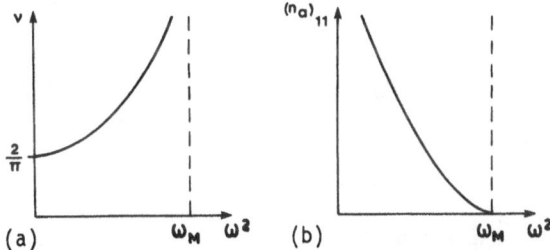

Fig.5.8. Comparison of the densities of states of a perfect linear chain (a) and of a linear chain containing a vacancy (b)

The suppression of the force constants between the central atom and its neighbors leads to a decrease of the diagonal terms on these atoms and consequently to a decrease of their frequencies. This conclusion can be extended to more complex cases.

5.3 Experimental Determination of Vibrational Modes

The measurements of localized vibrational modes associated with a defect are a powerful method for the identification of impurities or impurity related defects. In the simple case of a Group III or V substitutional impurity in a Group IV semiconductor, the change in the force constants can be neglected and the vibrational mode ω_E, in the Einstein model, is simply related to the mass M of the impurity (see Sect.5.2.3) through (5.82):

$$\omega_E = \omega_{E0} \sqrt{\frac{M_0}{M}} \, , \tag{5.103}$$

or

$$\omega_E = \omega_M \sqrt{\frac{M_0}{2M}} \, , \tag{5.104}$$

if we take ω_M equal to $\omega_{E0}\sqrt{2}$, as in true in the simple model of Sect.5.1.2. For the isotopes 10 and 11 of boron in silicon ($M_0 = 28$, $\hbar\omega_M = 0.0647$ eV), the localized modes ω_1 obtained with this expression compare favorably, as shown in Table 5.1, with the experimental data. In the case of a more complex defect involving one or more impurities, the identification of these impurities can be done when they have at least two isotopes. Since for two isotopes (M and M') of the same impurity there are the same changes in the force constants, the associated vibrational modes ω_E' and ω_E are such that:

$$\frac{\omega_E^2}{\omega_E'^2} = \frac{M'}{M} \, . \tag{5.105}$$

Table 5.1. Calculated (using expression 5.104) and experimental values [eV] (from [5.9]) of the vibrational modes associated with the two isotopes of substitutional boron in silicon

Element	M	$\hbar\omega_E$ calculated	$\hbar\omega_E$ experimental
B^{10}	10	0.0765	0.0802
B^{11}	11	0.0730	0.0772

For the isotopes of boron (M = 10, M' = 11) the ratio ω_E/ω_E' obtained using this expression is 1.049 and the experimental one (see Table 5.1) is 1.039.

How are these localized modes detected? They correspond to vibrations having an energy of the order of $\hbar\omega_M$ or higher, typically 65 meV in silicon, i.e., corresponding to a wavelength of 20 μ or below. They can interact with an electromagnetic wave, through the electric field $\underline{E}$. We shall describe this interaction on a very simple model, the mass defect with no change in force constants, treated in the Einstein approximation. As we have seen before this description is valid for strongly localized modes.

To ensure that there is a coupling of the defect (impurity atom) with the electric field $\underline{E}$, we consider that this impurity bears a charge q (reduced to a point charge at the impurity site). For a neutral defect there must be a compensating charge-q distributed on its neighbors. The phonon Hamiltonian discussed in the first section must be modified by adding the energy of interaction of each atomic charge q_i with the electric field. Instead of deriving the quantum solutions of this problem, it is simpler to use a classical formulation which leads to the same results. Let us then write the equation of motion for the impurity atom located at site 0. We have (using the notations of Sect.5.1)

$$M \frac{d^2}{dt^2} u_{0\alpha} = - \sum_{j\beta} A_{0\alpha,j\beta} u_{j\beta} + q E_{0\alpha} \quad , \tag{5.106}$$

where the first term on the right-hand side corresponds to the force of (5.10). The search for a solution of the form $\exp(i\omega t)$ leads for the amplitudes of the displacements to

$$M\omega^2 u_{0\alpha} = \sum_{j\beta} A_{0\alpha,j\beta} u_{j\beta} - q E_{0\alpha} \quad . \tag{5.107}$$

To find the complete solution, it is necessary to write down all other equations for the $u_{i\alpha}$. However, for frequencies lying in the vicinity of a localized mode, it is enough to use the Einstein model which consists in the neglect of the nondiagonal terms in the force constants matrix $A_{i\alpha,\,j\beta}$. Equation (5.107) then becomes

$$M\omega^2 u_{0\alpha} = A_{0\alpha,0\alpha} u_{0\alpha} - q E_{0\alpha} \quad . \tag{5.108}$$

We take $\underline{E}$ to be along the x direction. We can rewrite (5.108) to obtain

$$u_{0x} = - \frac{qE}{M\left(\omega^2 - \omega_E^2\right)} \quad , \tag{5.109}$$

where ω_E is the localized mode frequency in this Einstein model. In the limit of strongly localized states all other atomic displacements are negligible. The displacement u_{0x} of the charge q corresponds to a dipole of moment qu_{0x} along the x direction. If there are N defects per unit volume, the polarization P (equal to the dipole moment per unit volume) is equal to:

$$P = Nqu_{0x} \quad . \tag{5.110}$$

The dielectric constant ε of the system is given by $1 + 4\pi P/E$. With the help of (5.109,110) it can be written

$$\varepsilon(\omega) = 1 + \frac{4\pi Nq^2}{M\left(\omega_E^2 - \omega^2\right)} \quad . \tag{5.111}$$

There are contributions to $\varepsilon(\omega)$ from the other states as well, but for the frequencies ω in the vicinity of ω_E, the contribution given by (5.111) clearly dominates. In this formula we introduce the static limit $\varepsilon(0)$ given by

$$\varepsilon(0) = 1 + \frac{4\pi Nq^2}{M\omega_E^2} \quad , \tag{5.112}$$

and the high frequency limit $\varepsilon(\infty)$ equal to unity. This allows $\varepsilon(\omega)$ to be rewritten under the well-known form:

$$\varepsilon(\omega) = \varepsilon(\infty) + [\varepsilon(\infty) - \varepsilon(0)] \frac{\omega_E^2}{\omega^2 - \omega_E^2} \quad . \tag{5.113}$$

This expression is derived in the harmonic approximation in which all the vibrational modes (phonons) are independent (the equations of motion are linear). Actually, the phonons are coupled through anharmonic terms, which means that part of the energy transmitted to one phonon mode by the electromagnetic field is transmitted to other phonon modes. In a first approximation this phonon coupling can be treated by the introduction of a term proportional to du/dt in the equation of motion. This is equivalent to introduce a damping term $i\Gamma\omega$ in the dielectric constant:

$$\varepsilon(\omega) = \varepsilon(\infty) + [\varepsilon(\infty) - \varepsilon(0)] \frac{\omega_E^2}{\omega^2 + i\Gamma\omega - \omega_E^2} \quad , \tag{5.114}$$

which then becomes a complex quantity,

$$\varepsilon = \varepsilon' + i\varepsilon'' \quad , \tag{5.115}$$

with

$$\varepsilon' = \varepsilon(\infty) + \frac{[\varepsilon(0)-\varepsilon(\infty)]\left(\omega_E^2-\omega^2\right)\omega_E^2}{\left(\omega_E^2-\omega^2\right)^2+\Gamma^2\omega^2} \qquad (5.116)$$

and

$$\varepsilon'' = \frac{[\varepsilon(0)-\varepsilon(\infty)]\omega_E^2\omega\Gamma}{\left(\omega_E^2-\omega^2\right)^2+\Gamma^2\omega^2} \qquad . \qquad (5.117)$$

Consider an electromagnetic plane wave of pulsation ω, propagating in the material having such a dielectric constant. We know that the initial amplitude a_0 of this wave is attenuated with the distance x of propagation in the following way:

$$a = a_0 \exp(-\omega kx/c) \quad , \qquad (5.118)$$

where c is the light velocity and k the imaginary part of the complex refractive index defined as

$$(n + ik)^2 = \varepsilon(\omega) \qquad . \qquad (5.119)$$

The real and imaginary parts of $\varepsilon(\omega)$ can be expressed in terms of n and k:

$$\varepsilon' = n^2 - k^2 \qquad (5.120)$$

$$\varepsilon'' = 2nk \quad , \qquad (5.121)$$

or inversely, n and k can be expressed in terms of ε' and ε'':

$$k = \left(\frac{\sqrt{\varepsilon'^2+\varepsilon''^2}-\varepsilon'}{2}\right)^{\frac{1}{2}} \qquad (5.122)$$

$$n = \frac{\varepsilon''^2}{2(\sqrt{\varepsilon'^2+\varepsilon''^2}-\varepsilon')} \qquad . \qquad (5.123)$$

As a result, the intensity of the light absorbed by a thickness x of the material,

$$|a|^2 = |a_0|^2 \exp(-2\omega kx/c) \quad , \qquad (5.124)$$

is maximum when ωk is maximum, i.e., when the common denominator of ε' and ε'' is minimum. This occurs when

$$\frac{d}{d\omega}\left[\left(\omega_E^2 - \omega^2\right)^2 + \Gamma^2\omega^2\right] = 0 \quad,$$

(5.125)

that is for

$$\omega = \left(\omega_E^2 - \frac{\Gamma}{2}\right)^{\frac{1}{2}} \quad.$$

(5.126)

This resonant pulsation actually differs very little from ω_E because, usually, $\Gamma/\omega_E \ll 1$. In the vicinity of the resonant frequency, the optical measurements which provide the maximum amount of information on the resonance parameters are reflectivity measurements at the normal incidence. Indeed, the reflectance power, ratio of the reflected to the incident light intensities, provide both the real and imaginary part of the refractive index.

Optical measurements on vibrational modes associated with defects are only possible when absorption by free carriers is negligible. The material studied must be practically intrinsic. This is obtained by the compensation of the doping impurity resulting from the introduction of deep levels by irradiation or by diffusion of shallow impurities of a different type than the doping impurity. This compensation introduces additional vibrational modes which complicate the measurements.

5.4 Vibrational Entropy

In this section we assume that the localized vibrational modes around a defect are known and calculate the corresponding vibrational entropy. We first derive the general expression of the contribution of vibrational modes to the entropy, then apply it to the case of the vacancy and show that, when atomic rearrangements around the vacancy are neglected, the formation entropy of this defect is about 1.5 k. We finally introduce lattice rearrangement in the calculation of the entropy and find that this contribution can increase it in very large proportions.

5.4.1 General Expression

We start from the thermodynamic expressions of the free energy F and of the entropy S, which are given by [5.9]

$$F = - kT \log Z$$

(5.127)

$$S = -(\partial F/\partial T)_V \quad,$$

(5.128)

where Z is the partition function, v the volume, and T the temperature. For simplicity, let us consider the case of a simple harmonic oscillator of frequency ω. Its partition function is given by [5.9]

$$Z = \frac{\exp(-x/2)}{1 - \exp(-x)} \quad , \tag{5.129}$$

where

$$x = \frac{\hbar\omega}{kT} \quad . \tag{5.130}$$

In the high-temperature limit, we can expand Z to second order in x. The same development for F gives

$$F = \frac{\hbar\omega}{2} + kT \log x - \frac{x}{2} kT \left(1 - \frac{x}{12}\right) \quad , \tag{5.131}$$

whereas we obtain for S

$$S = k\left(1 - \log x + \frac{x^2}{24}\right) \quad . \tag{5.132}$$

This development can be extended to complex systems, when they are treated in the harmonic approximation, since we have seen in the first part of this chapter that such systems can be considered as a superposition of 3 N harmonic oscillators (N is the number of atoms, each one having three independent displacements). Then, because the entropy is an additive function, we have

$$\frac{S}{k} = \sum_{\alpha} \left[1 - \log \frac{\hbar\omega_\alpha}{kT} + \frac{1}{24} \left(\frac{\hbar\omega_\alpha}{kT}\right)^2 \cdots\right] \quad . \tag{5.133}$$

In this expression the ω_α are the 3 N eigenfrequencies. However, this expansion is only useful at high temperatures. We could easily derive the general expression

$$\frac{S}{k} = \sum_{\alpha} \left\{\frac{\hbar\omega_\alpha/kT}{\exp(\hbar\omega_\alpha/kT)-1} - \log[1 - [\exp(-\hbar\omega_\alpha/kT)]\right\} \quad , \tag{5.134a}$$

which has the correct limiting behavior at low temperatures ($S \to 0$ as $T \to 0$). In the following, we develop simple applications and make use mainly of the expansion to second order in the high-temperature range. We also make use of a moment expansion which we now develop.

5.4.2 Expansion in Moments

An exact calculation of the entropy from (5.134a) requires, in general, heavy numerical computations. We shall see later that, once again, the Green's functions' method is the appropriate technique to obtain exact results in the case of defects. However, we can readily obtain quite good approximations by using an expansion technique, which involves the moments of the densities of states.

The method is based on the fact that the ω_α^2 are the eigenvalues of the dynamical matrix D. For instance, the high-temperature expansion of the entropy can be written as

$$\frac{S}{k} \simeq 3N(1 - \log \hbar/kT) - \frac{1}{2} \text{Trace}\{\log D\} \quad , \tag{5.134b}$$

where second-order terms have been dropped. To calculate this quantity, it is advantageous to separate D into its diagonal part d and its nondiagonal part R and to use the following relations:

$$\log D = \log[d(1 + d^{-1}R)] \quad , \tag{5.135}$$

that is

$$\log D = \log d + \sum_n \frac{(-1)^n}{n} (d^{-1}R)^n \quad . \tag{5.136}$$

As long as the expansion converges rapidly, we can write the entropy as

$$\frac{S}{K} \simeq 3N\left(1 - \log \frac{\hbar}{kT}\right) - \frac{1}{2} \text{Trace} \left\{\log d + \sum_n \frac{(-1)^n}{n} (d^{-1}R)^n\right\} \quad . \tag{5.137}$$

It has been shown [5.10] that such an expansion converges rapidly in most cases. We can even obtain meaningful information from consideration of the term log d alone. This approximation corresponds to the Einstein model, since only the diagonal part of the dynamical matrix is kept.

5.4.3 The Vacancy

As a simple illustration, let us now investigate the case of the formation entropy of the vacancy S_F [5.11]. The physical situation corresponds to the removal of an atom from its normal lattice site, putting it at a kink on the surface. As shown on Fig.6.1, this is strictly equivalent to increasing the number of volume sites by one. The total number of atoms N in the system does not change, so that the first term of (5.137), i.e., $3N(1 - \log \hbar/kT)$, does not contribute to the change in the entropy S_F due to vacancy formation. In the Einstein approximation, S_F can thus be written as

$$S_F \simeq - \frac{1}{2} \text{Trace} \{\log d'\} + \frac{1}{2} \text{Trace} \{\log d\} \quad , \tag{5.138}$$

or

$$S_F \simeq \frac{1}{2} \sum_{i,\alpha} \log \frac{d_{i\alpha,i\alpha}}{d'_{i\alpha,i\alpha}} \quad , \tag{5.139}$$

where d and d' are the diagonal parts of the unperturbed and perturbed systems. However, we must notice that, during vacancy formation, the perfect crystal of N atoms has been replaced by a system of N + 1 atoms plus a vacancy. The diagonal term which existed at the vacancy site (corresponding to the volume atom) has been regained by putting the corresponding atom on the surface (which is equivalent to an extra volume atom). Thus, the summation over i excludes the site i = 0 where the vacancy is created.

We now evaluate S_F in simple force constants models. One of these models assumes isotropic interactions between nearest neighbors, i.e.,

$$D_{i\alpha,j\beta} = - \frac{C}{M} \delta_{\alpha,\beta} \quad , \tag{5.140}$$

for any pair of nearest neighbors i and j (we take here a system of identical atoms with mass M). Translational invariance implies for the diagonal terms $D_{i\alpha,i\alpha}$ that

$$D_{i\alpha,i\alpha} = Z_i \frac{C}{M} \delta_{\alpha,\beta} \quad , \tag{5.141}$$

where Z_i is the number of nearest neighbors of atom i in the system. In this simple model only the nearest neigbors contribute to S_F, leading to

$$\frac{S_F}{k} \simeq \frac{3}{2} Z_0 \log \frac{Z_0}{Z_0 - 1} \quad , \tag{5.142}$$

where Z_0 is the coordination number in the perfect crystal. We can go still further by expanding this expression in powers of $1/Z_0$. The most important zeroth order term is independent of Z_0 and leads to the value

$$S_F \simeq 1.5 \, k \quad , \tag{5.143}$$

irrespective of the system. This is a very important result, giving a universal order of magnitude for S_F. This value is in reasonable agreement with experimental data for practically all materials except for semiconductors (see Sect.5.4.6).

In order to show that the value 1.5 k for S_F does not depend too much upon the model, let us consider now the case of central forces between nearest neighbors. In this situation, we obtain

$$D_{i\alpha,j\beta} = -\frac{C}{M}\frac{X_{ij}^{\alpha}X_{ij}^{\beta}}{R^2} \quad , \tag{5.144}$$

where X_{ij}^{α} is the αth component of the unit vector $\hat{r}_{ij}$, as in Sect.5.1.4. Then,

$$\frac{S_F}{k} \simeq \frac{1}{2}\sum_{i,\alpha} \log \frac{\sum_{j}|X_{ij}^{\alpha}|^2}{\sum_{j}|X_{ij}^{\alpha}|^2 - |X_{0i}|^2} \quad , \tag{5.145}$$

where i are the nearest neighbors of the vacancy (situated at site 0) while j are the nearest neigbors of i. Expanding in powers of $|X_{0i}^{\alpha}|^2/\sum_{j}|X_{ij}^{\alpha}|^2$, we again obtain

$$\frac{S_F}{k} \simeq \frac{1}{2}\sum_{i,\alpha} \frac{|X_{0i}^{\alpha}|^2}{\sum_{j}|X_{ij}^{\alpha}|^2} = 3/2 \quad . \tag{5.146}$$

Similar considerations apply to the first corrective term in $1/T^2$ in (5.133). This term involves the second moment $\sum_{\alpha} \omega_{\alpha}^2$ of the frequency distribution, which is just the trace of the dynamical matrix, i.e., the sum of its diagonal elements. This correction is then obtained exactly in the Einstein model.

5.4.4 The Vacancy in Covalent Materials

We investigate here the changes in entropy brought by atomic displacements (relaxation or distorsion). In semiconductors it is known, from electron paramagnetic resonance studies in particular, that there can be strong atomic rearrangements around defects. We shall see that the additional entropy term, introduced by the rearrangements in the case of the vacancy, can be very high for the amplitude of the relaxation expected in Group IV semiconductors.

The variation of entropy ΔS in self-diffusion is the sum of two terms: S_F, related to the formation of the vacancy, and S_m, related to its migration (see Sect.7.1),

$$\Delta S = S_F + S_m \quad . \tag{5.147}$$

We now calculate this change in entropy, including changes in force constants originating from relaxation effects.

a) *Entropy of Formation*

The removal of an atom from its substitutional position to create a vacancy can induce a relaxation of the surrounding lattice, which in turn induces a change in the corresponding force constants. We consider the case of Fig.5.9 in which the relaxation is limited to the first neighbors of the vacancy, the force constants changing from C_0 to C_1. The diagonal elements of the dynamical matrix then change from

$$d_{i\alpha,i\alpha} = 4 C_0/M \quad \text{to} \quad d'_{i\alpha,i\alpha} = 3 C_1/M \qquad (5.148)$$

for the nearest neighbors of the vacancy, and to

$$d'_{i\alpha,i\alpha} = (3 C_0 + C_1)/M \qquad (5.149)$$

for the next nearest neighbors. Consequently,

$$\sum_{i,\alpha} \ln \left(\frac{d_{i\alpha,i\alpha}}{d'_{i\alpha,i\alpha}} \right) = 12 \ln \frac{4}{3} + 12 \ln \frac{C_0}{C_1} + 36 \ln \left(\frac{4C_0}{3C_0+C_1} \right) , \qquad (5.150)$$

and

$$\sum_{i,\alpha} (d'_{i\alpha,i\alpha} - d_{i\alpha,i\alpha}) = \omega^2 \left[12 \left(\frac{3C_1}{4C_0} - 1 \right) + 36 \left(\frac{3C_0+C_1}{4C_0} - 1 \right) \right] . \qquad (5.151)$$

Thus,

$$\frac{\Delta S^f}{k} = 6 \ln \frac{4}{3} + 6 \ln \frac{C_0}{C_1} + 18 \ln \left(\frac{4C_0}{3C_0+C_1} \right) \qquad (5.152)$$

$$+ \left(\frac{\hbar\omega}{kT} \right)^2 \left[\frac{1}{2} \left(\frac{3C_1}{4C_0} - 1 \right) + \frac{3}{2} \left(\frac{3C_0+C_1}{4C_0} \right) - 1 \right) \right] , \quad \text{with}$$

$$\omega^2 = \frac{4C_0}{M} . \qquad (5.153)$$

b) *Entropy of Migration*

The entropy of migration ΔS_m corresponds to the entropy difference betweeen two states, one in which the vacancy is in its equilibrium position, the other in which the vacancy is in its saddle point. For the migrating atom, only two displacements (x,y) perpendicular to the migrating path (z) have to be considered (see Fig.5.10).

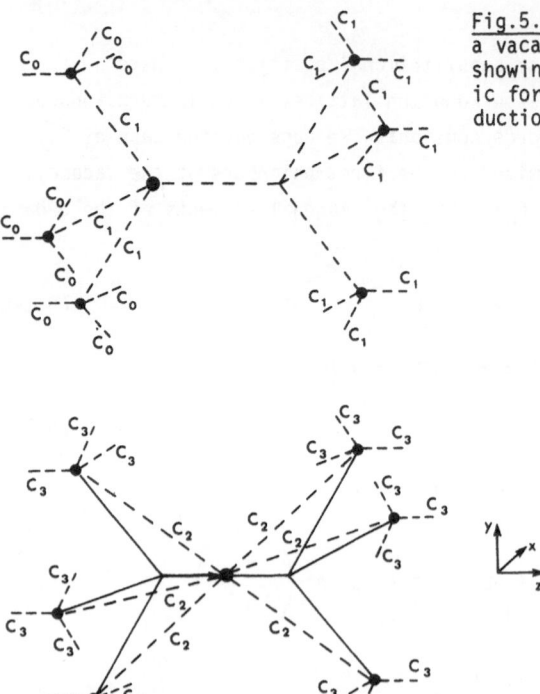

Fig.5.9. Lattice configuration around a vacancy in its equilibrium position showing the changes in the interatomic force constants due to the introduction of the vacancy in the lattice

Fig.5.10. Lattice configuration around a vacancy in its saddle position showing the various interatomic force constants involved

In the saddle point position (Fig.5.10), we assume that the force constant between the six neighbors of the migrating atom can be neglected. The force constant between this atom and its neighbors is now taken to be C_2. We also assume that the force constant between these six atoms and their own nearest neighbors remains C_1 in this position (these atoms have only three nearest neighbors as was the case for those of the vacancy in its equilibrium position of Fig.5.9). Then, considering separately the contributions to the entropy, we have

$$\frac{\Delta S_m}{k} = \ln\left(\frac{C_1}{2C_2}\right) + \frac{9}{2}\ln\left(\frac{3C_1}{3C_1+C_2}\right) + \frac{9}{2}\ln\left(\frac{3C_0+C_1}{3C_1+C_2}\right)$$

$$+ \frac{27}{2}\ln\left(\frac{4C_0}{3C_0+C_1}\right) + \frac{1}{32}\left(\frac{\hbar\omega}{kT}\right)^2\left(\frac{10C_2}{C_0} + \frac{13C_1}{C_0} - 18\right) \quad . \quad (5.154)$$

From the derived expressions it is possible to understand that a) at high temperature the value of ΔS can be considerably larger than the value usually expected (1 to 5 k), and b) at low temperature the entropy for migration has the usual value. Consider first the temperature independent term ΔS_0 in ΔS. The force constant C_2 is neglected in front of C_0 and C_1, and ΔS_0 can be written

$$\frac{\Delta S_0}{k} = 1.73 + \ln \frac{C_1}{2C_2} + \frac{63}{2} \ln \left(\frac{4C_0}{3C_0+C_1}\right) + 6 \ln \frac{C_0}{C_1} + \frac{9}{2} \ln \left(\frac{3C_0+C_1}{3C_1}\right) \quad . \tag{5.155}$$

A lower limit of this quantity can be obtained versus C_0/C_1 by dropping the quantity $\ln(C_1/2\,C_2)$. The corresponding variation of ΔS_0 is then given by Fig.5.11.

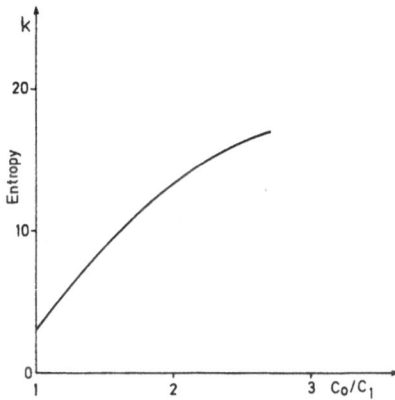

Fig.5.11. Variation of the entropy for self-diffusion versus the change in the force constants (from C_0 to C_1)

The force constants around a vacancy are unknown. However, calculations [5.12,13] performed using the techniques described in Chap.4 suggest that the first neighbors of a vacancy in diamond undergo an inward relaxation of about 10% of the interatomic distance. Then, using a Morse potential (Chap.6) to represent the bond energy $U(r)$ versus the interatomic distance r, we can compute the force constants,

$$C = \frac{\partial^2 U}{\partial r^2} \quad , \tag{5.156}$$

for the interatomic distance (C_0) and for a decrease of 10% (C_1), and obtain C_0/C_1. In the case of silicon, we obtain $C_0/C_1 \simeq 2$; the lower limit of ΔS_0 is then (see Fig.5.11) $\Delta S_0 \simeq 15$ k.

For $C_0/C_1 = 2$, the temperature dependent term ΔS_T in ΔS is

$$\frac{\Delta S_T}{k} = - 0.86 \left(\frac{\hbar\omega}{kT}\right)^2 \quad . \tag{5.157}$$

Taking $\omega = \omega_M/\sqrt{2}$, where ω_M is the Raman frequency ($\hbar\omega_M \simeq 0.06$ eV in silicon), we obtain

$$\frac{\Delta S_T}{k} = \frac{-2 \times 10^5}{T^2} \quad ; \tag{5.158}$$

where T is expressed in K. To detect a change of 1 k in ΔS, it is necessary to have experimental data down to $1000/T \simeq 2$, i.e., $T \simeq 500$ K. Unfortunately, there are no experimental data in this low temperature range (Chap.7).

5.4.5 Exact Calculation of Entropy Changes

The change in entropy induced by a point defect can be calculated exactly using the Green's functions techniques, as introduced in Chap.3 for the electronic structure. For vibrations the matrix to be diagonalized is $\dot{D}$ instead of H and its eigenvalues are ω^2 instead of the energy E. A simple transposition leads to the following definition of the resolvent G_0 of the perfect crystal:

$$G_0 = \left(\omega^2 - D_0\right)^{-1} \quad , \tag{5.159}$$

where D_0 is the dynamical matrix.

In many cases we can consider that the point defect leads to a localized perturbation V on the dynamical matrix. It is then of interest to calculate the change $\delta N(\omega^2)$ of the total number of states of squared frequency smaller than ω^2, under the form:

$$\delta N(\omega^2) = -\frac{1}{\pi} \tan^{-1} \frac{\text{Im}\{\det(I - G_0 V)\}}{\text{Re}\{\det(I - G_0 V)\}} \quad , \tag{5.160}$$

the determinant having the same size as the perturbation V itself (all this is detailed in Chap.3). Once the quantity $\delta N(\omega^2)$ is calculated, it is quite trivial to obtain any integral quantity such as ΔS since (5.134) can be rewritten as

$$S = \sum_\alpha f\left(\omega_\alpha^2\right) \quad , \tag{5.161}$$

that is,

$$S = \int f(\omega^2) n(\omega^2) d\omega^2 \quad , \tag{5.162}$$

where $n(\omega^2)$ is the density of states. Any change ΔS will involve the same expression, but with the change $\delta n(\omega^2)$ in density of states. Integrating this by parts we obtain

$$\Delta S = \int \frac{df(\omega^2)}{d\omega^2} \, \delta N(\omega^2) d\omega^2 \quad . \tag{5.163}$$

This expression is valid if there is no change in the total number of states in the system. It allows a complete numerical calculation of ΔS (which has been done in detail [5.14]) and gives a value still larger than in the Einstein model.

5.4.6 Experimental Determination of Entropies

Many physical quantities U related to a free energy $G = H - TS$ are thermally activated and can be written

$$W = W_0 \exp(S/k) \exp(-H/kT) \quad . \tag{5.164}$$

When W is a defect concentration, then S relates to a formation entropy; when W is the concentration of defects which anneal due to their migration, S relates to the migration entropy; when W is the rate of electron or hole emission from a defect state to the conduction or valence bands, S is an ionization entropy. The preexponential term, as measured by the extrapolation of the ln W versus T^{-1} plot to $T^{-1} = 0$, provides a measure of S. Of course, this preexponential term contains also other parameters. These parameters, which will be discussed in the corresponding chapters for various cases of interest, have to be measured independently.

As an illustration we consider the case of self-diffusion. The case of carrier emission from an electron trap leading to an entropy of ionization will be treated in Vol.2, Chap.12. The variation of ln D versus T^{-1}, where D is the self-diffusion coefficient, provides a value for the preexponential factor D_0 varying from 10^2 to 10^4 $cm^2 s^{-1}$ in silicon.

As we shall see in Chap.7, self-diffusion occurs through a vacancy mechanism and S is then the sum of the formation and migration entropies of the vacancy, i.e., is given by (5.147). The expression which relates S to D_0 (justified in Chap.7) is

$$S = k \ln \frac{D_0}{\nu a^2} \quad , \tag{5.165}$$

where a is the lattice parameter and ν the vibrational frequency. For silicon, $a = 4/\sqrt{3} \times 2.35$ Å and $\nu \simeq 1.3 \times 10^{13}$ s^{-1}. Consequently, S ranges from 8 to 16 k. This value is very large compared to the values observed experimentally in metals and ionic materials (where S varies from 1 to 3 k) which are in reasonable agreement with the theoretical ones ($S \simeq 1.5$ k). This led people to suggest that the self-diffusion is not occuring through a simple vacancy mechanism (Sect.7.4.3). However, as we have shown in Sect.5.4.4, such a high value can be obtained from a calculation which takes into account a reasonable lattice relaxation around the vacancy. The reason the entropy of the vacancy is larger in semiconductors than in other materials would then be that the distortion involved is, as expected, larger in semiconductors than in metals and ionic materials.

6. Thermodynamics of Defects

The aim of this chapter is to examine the nature and the concentration of
the intrinsic defects which exist at thermal equilibrium in a covalent material. The concentration of a given point defect, at a given temperature,
is a function of its free energy of formation. Therefore, we shall first
describe methods for obtaining the formation enthalpy of simple defects,
the formation entropy having been previously discussed in Chap.5[1]. Then, we
shall derive an expression for the defect concentration as a function of
temperature and, finally, discuss the nature of the defects that are present
in silicon and germanium.

6.1 Enthalpy of Formation

In order to form a vacancy, an atom is removed from its substitutional site
and placed on the surface. As illustrated in Fig.6.1, this is equivalent to
the creation of a new atomic site in the volume. The formation enthalpy of
a vacancy H_F^V (at constant pressure) is the energy required to do this. Similarly, the formation enthalpy of an interstitial H_F^I is the energy required
to take an atom from the surface and put it in the chosen interstitial site.
We can easily see that the formation enthalpy of a Frenkel (vacancy-interstitial) pair is $H_F^V + H_F^I$, if we assume that there is no interaction between
the vacancy and the interstitial.

In covalent materials, the theoretical estimates of formation enthalpies
fall into two categories. One is classical in nature; it relates the formation enthalpy to the bond dissociation energy and uses an empirical potential to describe the variation of the bond energy versus the interatomic distance. The second category corresponds to quantum-mechanical treatments
which follow two approaches: a one-electron energy-band approach and a

---

[1] The basis of the thermodynamics of solids and of the statistical interpretation of entropy can be found in [6.1]. The statistical thermodynamics
of defects in crystal is treated by [6.2].

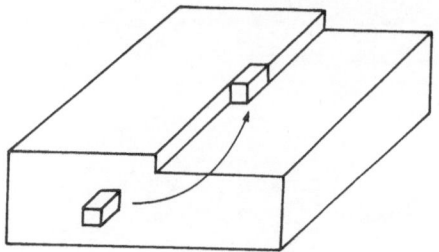

Fig.6.1. Process of formation of a vacancy; it consists of the removal of an atom from its lattice site and to place it at the surface, i.e., to add a new atomic volume to the solid

Table 6.1. Theoretical enthalpies H_F [eV] of formation for the vacancy (V) and the self-interstitial (I) in diamond, silicon, and germanium

	H_F^V	H_F^I	Ref.
Diamond	4.16		[6.3]
Silicon	2.32	-	[6.3]
	2.35	-	[6.4]
	2.13	0.93	[6.5]
Germanium	2.07	-	[6.3]
	1.97	-	[6.4]
	1.91	1.09	[6.5]
	2.21	-	[6.6]

molecular-orbital approach. The molecular-orbital approach is based on the concept of the defect molecule described in Chap.3. The band approach is based on a free-electron pseudopotential scattering by constructing a lattice potential of charged nuclei immersed in an electron gas. Several of these approaches have been applied to the vacancy but only one to the selfinterstitial. Considering the very crude approximations which are made in these calculations, the results can only be considered as estimates. Because the nature of the approximations is different from one approach to another, these approaches should not provide the same results. Indeed, the calculations lead to different conclusions; for instance, the classical approach demonstrates that the dominant term in the formation enthalpy is due to the lattice relaxation and distortion around the vacancy, a term which is neglected in the quantum-mechanical approaches. However, they surprisingly provide nearly the same results, as shown in Table 6.1.

Since the quantum treatments were briefly outlined in Chaps.3 and 4, we shall only develop in this section the classical treatment applied to the vacancy, as it was initially proposed by SWALIN [6.3]. This treatment assumes that the system can be considered as a superposition of strong covalent bonds. To take an atom out of its lattice site, four bonds are broken. To put this atom at the surface requires the formation of two new bonds. Then, if D is the bond dissociation energy, we have

$$H_F^V = 2 D \quad . \tag{6.1}$$

This bond energy can be evaluated from the sublimation (also called cohesive) energy E_c, which is the energy required to take an atom from the surface, i.e., to break two bonds, and send it to infinity

$$E_c = 2 D - E_{sp} \quad , \tag{6.2.}$$

where E_{sp} is the $s^2p^2 \rightarrow sp^3$ promotion energy (as recalled in Chap.3, in a covalent tetrahedral solid the electronic configuration corresponds to the promotion of an electron from an s to a p state). Table 6.2 gives the values of E_c, E_{sp}, and the values that are deduced for D and H_F^V in the case of Group IV elements.

Table 6.2. Cohesive E_c, promotion E_{sp}, deduced bond D, and vacancy formation H_F^V enthalpies [eV] in Group IV semiconductors

	E_c	E_{sp}	D	H_F^V
Diamond	7.8	8	7.9	15.8
Silicon	4.66	7.2	5.9	11.8
Germanium	3.85	8.1	6	12

A more realistic estimation should take into account the fact that new bonds are formed between the (broken) dangling bonds of the four neighbors of the vacancy, so that these neighbors undergo a relaxation and a distortion. Then,

$$H_F^V = 2 D - V_B + V_D \quad , \tag{6.3}$$

where V_B is the energy gained by forming these new bonds and V_D the energy associated with the deformation of all other bonds. Of course, these two energies are not independent, but we treat them separately in order to get their respective contributions. The evaluation of these energies necessitates the knowledge of the variation of the bond energy versus the interatomic distance r. A Morse potential,

$$V(r) = D[e^{-2\alpha(r-a)} - 2 e^{-\alpha(r-a)}] \tag{6.4}$$

is used that was originally derived for the interpretation of vibrational spectra of diatomic molecules. For covalent crystals, the parameters are

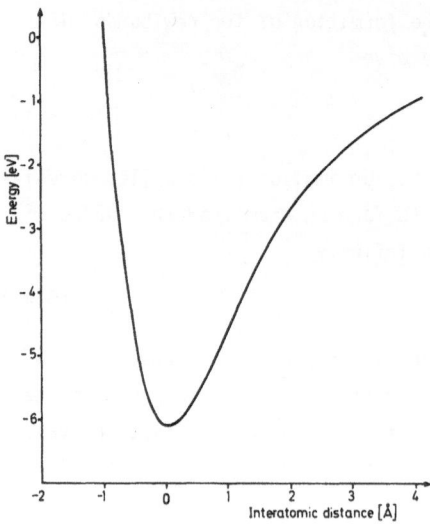

Fig.6.2. The Morse potential for germanium in its generalized form as proposed in [6.4] (the interatomic distance a is 2.45 Å)

Table 6.3. Dissociation D, distorsion V_D, bonding V_B, and formation H_F^V enthalpies [eV] in silicon and germanium according to [6.4]

	D	V_D	V_B	H_F^V
Ge	6	1.52	11.55	1.97
Si	5.9	1.80	11.25	2.35

obtained by a fit to the experimental results (the constant α is given from compressibility data [6.7]). As a result, the Morse potential (given in Fig. 6.2 for Ge) provides a reasonable representation of the bond energy near the equilibrium interatomic separation $r = a$ [$V_B(r)$ has a minimum equal to -D for this equilibrium separation]. But its use in the case of large displacements is more questionable (even under a "refined" form such as the one used by SEEGER and SWANSON [6.4]). The quantity $V_D(r) - V_B(r)$ is minimized to obtain the equilibrium configuration and finally H_F^V. The values of H_F^V obtained in this way are given in Table 6.3.

The validity of the interatomic potential used being questionable, the significance of the results is limited. This can be easily seen when we examine the values of the different energy terms, given in Table 6.3, which are obtained for the equilibrium configuration (following SEEGER and SWANSON [6.4]). These values correspond to the neutral vacancy V^0 since the four dangling bonds are paired two by two. We can see that since $V_B \simeq 2D$ and $H_F^V \simeq V_D$, H_F^V is due to the deformation of the backbonds. If we consider the positive vacancy V^+, two dangling bonds are paired as before, forming a two-electron bond. The two other dangling bonds are also paired, but the bond only contains one electron, and consequently, its energy is weaker by a factor of about two. Then, $V_B \simeq 1.5$ D, while V_D is only slightly modified. As a result, in silicon,

$$H_F^{V^+} \simeq 0.5 \ D + V_D \quad , \quad \text{i.e.,} \quad H_F^{V^+} \simeq 5 \ eV \quad .$$

Such a change in H_F^V is unrealistic since the energy involved in the $V^0 \rightarrow V^+$ transition will be shown later to be much smaller than this value (see Chap.8, Vol.II, on the negative H center formed by V^{++}, V^+, V^0). This shows that Swalin's model is strongly oversimplified.

6.2 Defect Concentration at Thermal Equilibrium

6.2.1 The Case of One Kind of Defects at Low Concentration

At a given temperature the free energy G of a crystal,

$$G = H - T S \quad , \tag{6.5}$$

is at a minimum. There are always intrinsic defects present in a crystal. Let us for the moment consider only one kind of defects and assume their total number to be n.

Their presence, although it increases the enthalpy H, increases also the entropy S. If H_0 is the enthalpy of a perfect crystal and H_F the formation enthalpy of isolated defects, then, in the limit where n is small so that each defect can be considered independently

$$H = H_0 + n \ H_F \tag{6.6}$$

The increase of entropy is due to the increased disorder. One part is the configurational disorder (over the possible sites) whose contribution is S_d. As we have seen in Chap.5, the other part, S_F, is associated with the disorder induced by the lattice vibrations. It can be calculated from the change in vibrational properties in the neighborhood of each isolated defect. The defect concentration is therefore obtained by minimizing the total change ΔG in free energy

$$\Delta G = G - G_0 = n(H_F - T \ S_F) - T \ S_d \quad , \tag{6.7}$$

(where G_0 is the free energy of the perfect crystal) i.e., by writing

$$\frac{\partial \Delta G}{\partial n} = G_F - T \frac{\partial S_d}{\partial n} \quad . \tag{6.8}$$

Here G_F, equal to $H_F - TS_F$, is the free energy of formation of a single isolated defect.

Let us calculate the entropy S_d due to the disorder. It is

$$S_d = k \ln(W) \quad , \tag{6.9}$$

where W, the complexion number, usually reduces to the number of distinct ways there are to distribute the n defects over the N possible sites

$$W = \frac{N(N - 1)(N - 2) \ldots (N - n + 1)}{n!} = \frac{N!}{(N - n)!n!} \quad , \tag{6.10}$$

where (!) stands for the factorial function. The quantities n and N being large, Stirling's formula

$$\log x! = x(\log x - 1) \quad , \tag{6.11}$$

can be applied, leading to

$$S_d = k\{N(\log N - 1) - (N - n)[\log(N - n) - 1] - n(\log n - 1)\} \quad . \tag{6.12}$$

In calculating $\partial S_d/\partial n$, care must be taken that N, the number of available sites for the defect, can depend upon n (this will be discussed in detail in the following). We find

$$\frac{\partial S_d}{\partial n} = k \left(\frac{\partial N}{\partial n} \log \frac{N}{N - n} + \log \frac{N - n}{n} \right) \quad . \tag{6.13}$$

There are many situations of interest in which the first term in (6.13) can be neglected:

    a) if N does not depend upon n,
    b) if n is much smaller than N, i.e., for low concentrations.

In the second case, (6.13) reduces to $k \log(N/n)$ and (6.8) leads to

$$G_F + k T \log \frac{n}{N} = 0 \quad , \tag{6.14}$$

or

$$\frac{n}{N} = \exp\left(-\frac{G_F}{kT} \right) \quad . \tag{6.15}$$

While (6.15) holds at low concentrations of defects, it is of interest to consider the possible departures. For this, we examine two intrinsic defects of great importance, the vacancy and the interstitial. These examples show also how N can be dependent on n.

Let us then consider a crystal containing N_0 identical atoms. Each time one atom is removed to create a vacancy, it is placed on a kink at the surface (Fig.6.1), which is equivalent to create a new atomic site. If there are n vacancies, the number of available sites is $N_0 + n$. Application of the exact formula (6.13), with $\partial N/\partial n$ equal to 1, leads to

$$\frac{n}{N_0 + n} = \exp\left(-\frac{G_F^V}{kT}\right) . \tag{6.16}$$

The interstitial case is somewhat analogous. We assume for the moment that there is one interstitial site per bulk atom. Any interstitial is considered to come from the surface, suppressing one bulk site, i.e., one interstitial site. For n interstitials, there are $(N_0 - n)$ available sites, leading to a value of -1 for $\partial N/\partial n$, and (6.13) gives

$$\frac{n(N_0 - n)}{(N_0 - 2n)^2} = \exp\left(-\frac{G_F^I}{kT}\right) . \tag{6.17}$$

Of course both expressions reduce to (6.15) at small concentrations.

6.2.2 Generalization to Several Kinds of Independent Defects and Internal Degrees of Freedom

In real situations there are presumably several kinds of defects. It is interesting to know if we can generalize in this case (6.15), obtained for the concentration of a given type of defect. For simplicity, we consider first that all the defects are independent. This assumption is valid if the defect potentials are short-ranged and becomes exact at very low concentrations. With this hypothesis, we can write the change ΔG in free energy as

$$\Delta G = \sum_i n_i G_{Fi} - kT \log(W) , \tag{6.18}$$

where n_i is the number defects of the ith kind, G_{Fi} the free energy of this isolated defect, and W the total complexion number which we now have to determine.

Let us first consider the trivial case where the defects do not occur on the same sites, so that there is no possibility of exclusion. This is true for vacancies and interstitials, if they do not interact even at their closest distance of approach. Then, the total complexion number W can be written as the product $\pi_i\ W_i$ of the W_i corresponding to each kind of defect. Because ΔG is additive and because the defects can be treated independently, (6.15) applies for each defect in the limit of low concentrations.

The problem is thus restricted now to defects which can occur on the same sort of sites. If n is the total number of defects, W can be written

$$W = \frac{N!}{(N-n)!\ \pi_i\ n_i!}\ .$$

(6.19)

The generalization of (6.13) is

$$\frac{\partial S_d}{\partial n_i} = k\left(\frac{\partial N}{\partial n_i}\ \log\frac{n}{N-n} + \log\frac{N-n}{n_i}\right)\ .$$

(6.20)

Unless N is independent of n_i (which is not true in general), the first term becomes vanishingly small only in the limit where $n \ll N$, i.e., for a small total concentration of defects. In this limit (6.20) reduces to $k\ \log(N/n_i)$, and we have

$$\frac{\partial \Delta G}{\partial n_i} = G_{Fi} + kT\ \log\frac{n_i}{N}\ ,$$

(6.21)

that is,

$$\frac{n_i}{N} = \exp\left(-\frac{G_{Fi}}{kT}\right)\ ,$$

(6.22)

which generalizes the result given by (6.15).

Another very important situation corresponds to the case where a defect has several internal degrees of freedom. There are many possible examples, for instance, when the defect can take several equivalent configurations, i.e., several orientations in the unit cell. This is true again when there is spin degeneracy or any other sort of degeneracy for the isolated defect. Let us call Z_i the number of equivalent possibilities. We can repeat all of the above arguments by replacing the number of available sites N by $Z_i N$. However, when one defect is in one of the Z_i possible configurations associated with a given site, it prevents any other defect from taking another of the $Z_i - 1$ remaining configurations at the same site. This simply multiplies W_i by $Z_i^{n_i}$, i.e., is equivalent to adding an entropy term $-k\ \log(Z_i)$ to G_{Fi}. In that case (6.22) has to be replaced by

$$\frac{n_i}{Z_i N} = \exp\left(-\frac{G_{Fi}}{kT}\right) \quad . \tag{6.23}$$

6.2.3 Equilibrium Between the Different Charge States of a Defect

In semiconductors, a given defect, say the vacancy, can exhibit different charge states. In Silicon, for instance, there are at least four reported charge states for the vacancy: V^+, V^0, V^- and V^{--}. Their relative concentrations at thermal equilibrium depend upon the Fermi-level position. Let us then study this dependence, once more in the low-concentration limit where their interactions become negligible (even for charged defects with a long-range Coulomb potential).

The problem of dealing with charged states of defects is complicated by the existence of an overall constraint corresponding to global charge neutrality. The condition of charge neutrality imposes the chemical potential of the electron, i.e., the Fermi level in Fermi-Dirac statistics. However, it is possible to calculate directly the concentration of charged defects and express it in terms of the electron chemical potential. The easiest way to achieve this is to consider a reaction that involves the charged defect and automatically satisfies the charge neutrality constraint. As an example, let us consider one possible reaction,

$$V^0 \rightleftharpoons V^+ + e^- \quad , \tag{6.24}$$

where the positive vacancy (V^+) is obtained by the ionization of an electron (e^-) from the neutral vacancy (V^0). Obviously, this reaction respects the neutrality constraint. The dissociated pairs $V^+ + e^-$ must be in equilibrium with the neutral defects V^0. The free energy G is a function of n_0 and n_+, the number of neutral and positive vacancies. At equilibrium, G must be minimum with respect to n_0 and n_+:

$$dG = \frac{\partial G}{\partial n_0} dn_0 + \frac{\partial G}{\partial n_+} dn_+ = 0 \quad . \tag{6.25}$$

The concentrations n_0 and n_+ being related through the constraint (6.24)

$$dn_0 + dn_+ = 0 \quad ; \tag{6.26}$$

we can thus write

$$\left(\frac{\partial G}{\partial n_0} - \frac{\partial G}{\partial n_+}\right) dn_0 = 0 \quad , \tag{6.27}$$

which means that the equilibrium condition dG = 0 is given by

$$\frac{\partial G}{\partial n_0} = \frac{\partial G}{\partial n_+} \quad . \tag{6.28}$$

This relation expresses the equality of the chemical potentials.

For noninteracting defects we can use (6.21) and write

$$\frac{\partial G}{\partial n_0} = G_F(V^0) + kT \log \frac{n_0}{N_0} \quad , \tag{6.29}$$

and

$$\frac{\partial G}{\partial n_+} = \frac{\partial G(V^+)}{\partial n_+} + \frac{\partial G(e^-)}{\partial n_+} \quad , \tag{6.30}$$

that is,

$$\frac{\partial G}{\partial n_+} = G_F(V^+) + kT \log \frac{n_+}{N_+} + \mu_{e^-} \quad . \tag{6.31}$$

In these expressions, $G_F(V^0)$ and $G_F(V^+)$ are the free energies of formation of the isolated vacancies V^0 and V^+; N_0 and N_+ are the number of available sites for the vacancies (including possible internal degeneracies). To derive (6.30), we have used the fact that the free energy of the separated pairs (V^+, e^-) is the sum of the free energies $G(V^+)$ and $G(e^-)$ of each component of the pair. We have also used the fact that $\partial G(e^-)/\partial n_+$ is by definition the chemical potential μ_{e^-} of the electron i.e., the Fermi level E_F in Fermi-Dirac statistics. From (6.28-31), we can finally write

$$\frac{n_+}{N_+} \frac{N_0}{n_0} = \exp\left[-\frac{G_F(V^+) - G_F(V^0) + E_F}{kT}\right] \quad . \tag{6.32}$$

Usually the concentrations of vacancies $c(V^0)$ and $c(V^+)$ are expressed as the ratios of n_0 and n_+ over the number of atomic sites. If we call $Z(V^0)$ and $Z(V^+)$ the internal degeneracies, we can reexpress (6.32) under the form

$$\frac{c(V^+)}{c(V^0)} = \frac{Z(V^+)}{Z(V^0)} \exp\left[\frac{E_t(V^0) - E_F}{kT}\right] \quad , \tag{6.33}$$

where E_t is equal to the difference $G(V^0) - G(V^+)$, i.e., is equal to the negative free energy of ionization of V^0. We leave the ratio $Z(V^+)/Z(V^0)$ unspecified since it can only be obtained by a detailed numerical calculation of the ground states of V^+ and V^0.

A symmetrical situation holds for V^0 and V^-. When writing the reaction

$$V^- \rightleftharpoons V^0 + e^- \ , \qquad\qquad (6.34)$$

we immediately obtain

$$\frac{c(V^-)}{c(V^0)} = \frac{Z(V^-)}{Z(V^0)} \exp\left[-\frac{E_t(V^-)-E_F}{kT} \right] \ . \qquad\qquad (6.35)$$

Equations (6.33,35) are completely general and allow the concentrations of the charged vacancies to be obtained in terms of $c(V^0)$ and of the electron chemical potential. The quantity $c(V^0)$ is directly fixed through the usual expression (6.23), while E_F is fixed by the charge neutrality condition.

6.2.4 Defects in Short-Range Interaction

The previous sections assume that there is no interaction between defects, even when they have different charge states. This is obviously incorrect. The interactions very much complicate the problem. However, we can still obtain simple solutions in some limiting situations. The first of these limiting situations corresponds to short-range interactions between defects in such a way that they can form small, stable aggregates which can again be treated independently. The simplest examples correspond to vacancies that associate to form divacancies, and vacancies and interstitials that form vacancy-interstitial pairs. For such small aggregates we can repeat the previous treatment used for isolated defects. The only problem is to derive the available number of sites for each type of defect, as well as the corresponding internal degeneracy.

Let us begin with pairs of identical defects, for instance, the divacancy. In the low-concentration limit, we know that each different type of defect can be treated independently. Let us then call N the number of available atomic sites. There are thus N sites where we can fix one member of the pair. For each of these sites, we have to determine the degeneracy factor. If each atom has Z nearest neighbors, this degeneracy factor is equal to $Z/2$ because each pair corresponds to two neighboring sites (in a crystal there are $Z/2$ bonds per site). We thus end up with a number n_{2v} of divacancies equal to

$$\frac{n_{2v}}{N} = \frac{Z}{2} \exp\left(-\frac{G_F^{2v}}{kT} \right) \ . \qquad\qquad (6.36)$$

We can generalize the argument to any kind of complex defect as long as the different entities can be treated independently, i.e., in the low-concentration limit. The general formula will always be of the form

$$\frac{n_\ell}{N} = Z_\ell \exp\left(-\frac{G_{F\ell}}{kT}\right) , \qquad (6.37)$$

where Z_ℓ is the degeneracy factor of each defect per atomic site. We have seen that, for identical defects, Z_ℓ is equal to $Z/2$. For pairs of nonidentical defects, it becomes equal to Z, the permutation of the components of the pair leading to an extra degeneracy factor of two.

The free energy G_{Fp} for the formation of a pair of defects is often written under the convenient form

$$G_{Fp} = G_{F1} + G_{F2} - \Delta G_p , \qquad (6.38)$$

where G_{F1} and G_{F2} are the free energies of formation of the individual components, and ΔG_p is the binding free energy. The number p of pairs can be related to the product $n_1 \times n_2$ of individual defects at thermal equilibrium using relation (6.37)

$$p = \frac{n_1 n_2}{N} \frac{Z_p}{Z_1 Z_2} \exp\left(\frac{\Delta G_p}{kT}\right) \qquad (6.39)$$

where Z_p is the degeneracy factor of the pair.

The derivation of (6.39) is based on the fact that n_1, n_2, and p are independent numbers, which are calculated by minimizing the total free energy. This is not always the case, for instance, when considering pairing between two impurities, Here the concentration of each type of impurities is known and the only unknown is the number of pairs. Let us then consider this problem in more detail. We assume N_1 impurities of type 1, N_2 of type 2, and call p the number of pairs that occur only at nearest neighbor distances. The free energy change due to pair formation is given by

$$\Delta G = -p \, \Delta G_p - kT \log \frac{W}{W_0} , \qquad (6.40)$$

where W_0 and W are the complexion numbers before and after pairing. We have

$$W_0 = \frac{N!}{(N-N_1-N_2)! \, N_1! \, N_2!} , \qquad (6.41)$$

and

$$W = \frac{Z^P N!}{(N-n_1-n_2-p)! \, n_1! \, n_2! \, p!} \, , \qquad (6.42)$$

where Z is the degeneracy factor of the pairs (equal to the number of nearest equivalent sites in the simple case of different atoms); n_1 and n_2 are the numbers of atoms of type 1 and 2 which have not been paired. It is clear that n_1 and n_2 are equal to $N_1 - p$ and $N_2 - p$ respectively. If we use Stirling's formula and minimize G with respect to p, the final answer turns out to be

$$p = \frac{n_1 n_2}{(N+p)} \, Z \, \exp\left(\frac{\Delta G_p}{kT}\right) . \qquad (6.43)$$

In the limit where $p \ll N$, this expression takes exactly the same form as (6.39) but, here, care must be taken that n_1 and n_2 are respectively equal to $N_1 - p$ and $N_2 - p$.

6.2.5 The Case of Long-Range Interaction

The long-range interaction between defects can be of elastic or electrostatic origin. Because of the relaxation and distortion of the lattice around a defect, there is an associated strain energy. Such a strain energy can be reduced by pairing: a defect that induces a lattice compression will tend to pair with a defect that induces a lattice expansion. The elastic interaction varies as r^{-3}, where r is the distance between the two defects [6.8][2]. The magnitude of this interaction is not known in covalent materials. But if we assume it is similar to the metal case (i.e., corresponds to a critical radius of 8 Å at 80 K), then its variation is given in Fig.6.3.

Coulomb interaction can also result in defect pairing when the two defects possess charges (q_1 and q_2) of opposite sign, the gain in energy ΔE being

$$\Delta E = \frac{|q_1 q_2|}{\varepsilon r} \, , \qquad (6.44)$$

ε is the dielectric constant. This Coulomb attraction occurs over a rather large distance r. Consider Fig.6.3 in which the interaction $e^2/\varepsilon r$ is plotted versus r for silicon ($\Delta E[eV] = 0.87/r$ Å). The capture radius r_c, defined as the value of r for which the interaction energy becomes larger than the

---

[2] The continuum theory of lattice defects is not treated in this book. The theory of elastic interaction between point defects is developed in [6.9].

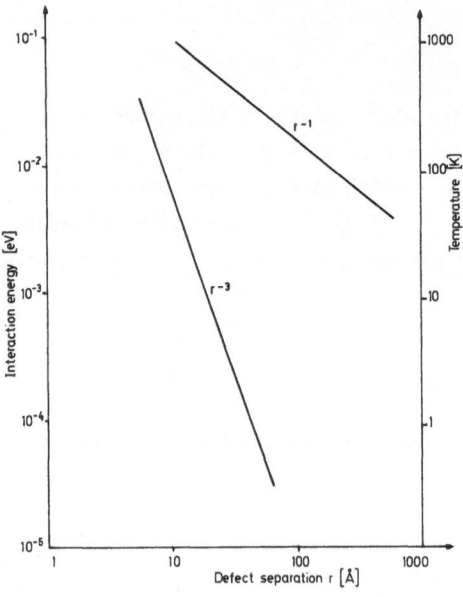

Fig.6.3. Variation of the elastic (r^{-3}) and Coulomb (r^{-1}) interactions between defects versus their separation r. The Coulomb interaction is given for silicon

thermal energy (kT), is considerably greater for the electrostatic interaction than for the elastic interaction. Typically, at 100 K the capture radius is $r_c \simeq 100$ Å for the electrostatic interaction and $r_c \simeq 7$ Å for the elastic interaction. Note that at low temperature, the Coulomb capture radius is very large (800 Å at 10 K). Consequently, for charged defects which are mobile at low temperature (the self-interstitial or the vacancy in silicon or germanium for instance), the electrostatic interaction should be an important factor in the annealing process.

With such a long-range interaction, the equilibrium concentrations of defects become more difficult to calculate. One possible method is again to consider the possible pairs of interacting defects 1 and 2 as independent entities. However, it is necessary to consider pairs occuring at all possible distances R_ℓ. The binding free energy of a pair at R_ℓ will be ΔG_ℓ. We can generalize the treatment described by (6.40-43), by calling p_ℓ the number of pairs at distance R_ℓ. We can write

$$\Delta G = - \sum_\ell p_\ell \, \Delta G_\ell - kT \log \frac{W}{W_0} \; , \tag{6.45}$$

with

$$W = \frac{N!}{(N-n_1-n_2 - \sum_\ell p_\ell)! \, n_1! \, n_2!} \; \prod_\ell \frac{Z_\ell^{p_\ell}}{p_\ell!} \; , \tag{6.46}$$

where

$$n_1 = N_1 - \sum_\ell P_\ell \qquad\qquad (6.47)$$

$$n_2 = N_2 - \sum_\ell P_\ell \quad . \qquad\qquad (6.48)$$

The minimization of ΔG with respect to each p_ℓ in these equations leads for each of them to a formula analogous to (6.43), i.e., to

$$P_\ell = Z_\ell \, \frac{n_1 n_2}{N} \, \exp\!\left(\frac{\Delta G_\ell}{kT}\right) \quad , \qquad\qquad (6.49)$$

valid if $\sum_\ell P_\ell \ll N$.

The arguments leading to (6.49) are not correct for the general case. They are meaningful only when associations between the pairs dominate, i.e., when other associations are much less probable. This means that, on the average, the interaction between the partners of a pair must be much greater than with other defects. This only occurs at low concentration and low temperature. Equation (6.49) also remains valid when N_1 and N_2 are not constant. If defects 1 and 2 are oppositely charged vacancies and interstitials, then n_1 and n_2 are now the thermal numbers of isolated defects, and ΔG_ℓ is the binding energy of a pair with respect to the isolated pair.

When this simple theory of long-range interaction is not valid, we must use another type of approximation derived from the Debye-Hückel theory of electrolytes [6.10]. For this we shall use a linear-response theory. By definition, the free energy G_i of an isolated defect of the ith type is the change in the free energy of the whole system caused by the introduction of one such defect at a given position $\vec{R}_i$. If the defect bears a charge q_i, then G_i can be written

$$G_i = G_{i0} + \delta G_i \quad , \qquad\qquad (6.50)$$

where G_{i0} corresponds to the neutral defect and δG_i to the introduction of the charge q_i at the defect site. Let us imagine a progressive charging process and consider an intermediate step where the charge is q. It creates a long-range perturbation that induces a change of the charge density $\delta\rho(q,\underline{r})$ of the system. This charge density creates an electrostatic potential $\delta\Psi(q,\underline{r})$ which is a function of q and of the position $\underline{r}$. For an infinitesimal amount of charge dq, the corresponding change dG_i in the free energy is equal to $\delta\Psi(q,\underline{r})$ dq and we can write

$$\delta G_i = \int_0^{q_i} \delta\Psi(q,\underline{R}_i) \, dq \quad . \tag{6.51}$$

If we assume q_i small enough so that the linear-response theory is applicable, then $\delta\Psi(q,\underline{r})$ can be expanded in powers of q; we retain only the first-order term, i.e.,

$$\delta\Psi(q,\underline{r}) = \alpha(\underline{r}) \, q \quad . \tag{6.52}$$

From this expression, we can integrate δG_i with respect to q and obtain

$$\delta G_i = \alpha(\underline{R}_i) \, \frac{q_i^2}{2} \quad , \tag{6.53}$$

that is,

$$\delta G_i = \frac{q_i \delta\Psi(\underline{R}_i)}{2} \quad . \tag{6.54}$$

Applying the considerations of the previous sections [see (6.22)], we obtain for the total number n_i of the defects of the ith kind

$$n_i = n_{i0} \, \exp\left(- \frac{q_i \delta\Psi(\underline{R}_i)}{2kT}\right) \quad , \tag{6.55}$$

where n_{i0} is the number we would obtain in the absence of electrostatic effects.

The problem is now to calculate the potential $\delta\Psi(\underline{r})$. For this, we consider around $\underline{R}_i$, a sphere of radius R which is the distance of closest approach of the other defects. Outside this sphere, the potential $\Psi(\underline{r})$ is the sum of the direct potential $\Psi(\underline{r})$ due to the charge q_i and of $\delta\Psi(\underline{r})$. A central approximation is to replace this microscopic potential by its macroscopic component $\bar{\Psi}(\underline{r})$ which is the average of $\Psi(\underline{r})$ over a volume V centered on the position $\underline{r}$ (this volume V must be large enough to contain many defects and small compared to the size of the whole system). $\bar{\Psi}(\underline{r})$ varies slowly in space and can be considered as constant over large regions. Due to the influence of $\bar{\Psi}(\underline{r})$, the concentration of the charged defects will vary in space. In the volume V around a given point $\underline{r}$, each defect bearing a charge q_j will have its free energy of formation modified by $q_j\bar{\Psi}(\underline{r})$. At equilibrium the number of these defects within V will be changed from n_j to $n_j + \delta n_j$, with

$$\delta n_j = n_j \left[\exp - \frac{q_j\bar{\Psi}(\underline{r})}{kT} - 1\right] \quad , \tag{6.56}$$

that is, to first order

$$\delta n_j \simeq - \frac{n_j q_j}{kT} \bar{\Psi}(\underline{r}) \quad . \tag{6.57}$$

This change in defect number induces in the volume V the following macro-scopic change in charge density:

$$\delta\bar{\rho}(r) = \frac{1}{V} \sum_j q_j \delta n_j \quad , \tag{6.58}$$

that is,

$$\delta\bar{\rho}(r) = - \frac{1}{V} \sum_j \frac{n_j q_j^2}{kT} \bar{\Psi}(\underline{r}) \quad , \tag{6.59}$$

the summation being performed on the charged defects j belonging to V. The potential $\bar{\Psi}(\underline{r})$ can be obtained from $\delta\bar{\rho}(\underline{r})$ by Poisson's equation which for $r > R$, leads to

$$\Delta\bar{\Psi}(\underline{r}) = - \frac{4\pi}{\varepsilon} \delta\bar{\rho}(\underline{r}) \quad , \tag{6.60}$$

that is

$$\Delta\bar{\Psi}(\underline{r}) = \frac{4\pi}{\varepsilon kT} \frac{1}{V} \sum_j n_j q_j^2 \bar{\Psi}(\underline{r}) \quad , \tag{6.61}$$

or

$$\Delta\bar{\Psi}(\underline{r}) = \frac{\bar{\Psi}(\underline{r})}{L_D^2} \quad . \tag{6.62}$$

Here ε is the dielectric constant of the medium and L_D, defined by (6.61,62), the associated Debye length. The problem has a spherical symmetry around R_i so that the equation reduces to its radial part. The general solution is then, for $r > R$,

$$\bar{\Psi}(r) = \frac{B}{r} \exp\left(- \frac{r}{L_D}\right) \tag{6.63}$$

where the unknown constant B must be determined by the continuity conditions at $r = R$. Inside the sphere, there is no charge density except q_i at the center. The total potential has thus the form $(q_i/r) + A$, where A is the value $\delta\Psi(R_i)$ which we want to determine. Since A and B are determined by the continuity of the potential and its derivative at $r = R$, we have

$$\frac{q_i}{R} + A = \frac{B}{R} \exp(-R/L_D) \quad , \tag{6.64}$$

and

$$\frac{-q_i}{R^2} = - \frac{B}{R^2}\left(1 + \frac{R}{L_D}\right)\exp(-R/L_D) \quad . \tag{6.65}$$

Solving this system, we obtain

$$\delta\Psi(\underline{R}_i) = A = \frac{-q_i}{R+L_D} \quad , \tag{6.66}$$

so that the final result for the atomic fraction n_i/N of any charged defect becomes

$$\frac{n_i}{N} = \frac{n_{i0}}{N}\exp\left(\frac{q_i^2}{2(R+L_D)kT}\right) \quad . \tag{6.67}$$

The final result is that the concentration of unassociated charged defects is enhanced by the inclusion of long-range Coulomb interaction. This is not true for neutral pairs. However, there is a problem in applying (6.67) because there is no precise definition for R (it is usually taken to be of the order of the interatomic distance). Fortunately in most cases, L_D is much larger than R so that the result is relatively insensitive to the value of R.

6.2.6 Defect Concentration in a Stoichiometric Compound

Here we consider a material composed of two types of atoms A and B, belonging to two different sublattices (a II-VI or a III-V compound). At a given temperature, this material contains a certain amount of vacancies or interstitials in each of the two sublattices, plus the possibility of antisite defects. Let n_a^v, n_b^v, n_a^i, n_b^i be the concentrations of vacancies and interstitials in both sublattices; we neglect the antisite defects. Since the compound is stoichiometric, we have

$$N_a - n_a^i + n_a^v = N_b - n_b^i + n_b^v \quad , \tag{6.68}$$

that is

$$n_a^v - n_a^i = n_b^v - n_b^i \quad , \tag{6.69}$$

because the number of sites in the two sublattices are equal:

$$N_a = N_b = N \quad . \tag{6.70}$$

Because the four quantities n_a^v, n_a^i, n_b^v, n_b^i are not independent, they cannot be determined directly by minimization of the free energy as in the previous sections. We have to include the supplementary constraint (6.69) into the minimization procedure. The general mathematical method to achieve this is the Lagrange multipliers technique. We shall follow here another method whose advantage is to lead to equations which are also valid in nonstoichiometric compounds.

The central idea is to consider the equilibrium between two kinds of different defects, imposing the stoichiometric constraint for the corresponding equilibrium. For any such pair (ℓ, j) it is not difficult to show that stoichiometry imposes one of the two following conditions:

$$d\, n_\ell + d\, n_j = 0 \quad , \tag{6.71}$$

or

$$d\, n_\ell - d\, n_j = 0 \quad . \tag{6.72}$$

Each equilibrium requires that G be an extremum, i.e.,

$$dG = \frac{\partial G}{\partial n_\ell}\, d\, n_\ell + \frac{\partial G}{\partial n_j}\, d\, n_j = 0 \quad , \tag{6.73}$$

that is,

$$dG = d\, n_\ell \left(\frac{\partial G}{\partial n_\ell} \pm \frac{\partial G}{\partial n_j} \right) = 0 \quad . \tag{6.74}$$

By using a formalism similar to that of Sect.6.2.3 and calling G_ℓ and G_j the individual free energies, then

$$\frac{n_\ell\, n_j}{N^2} = Z_\ell Z_j \exp\left(-\frac{G_\ell \pm G_j}{kT}\right) , \tag{6.75}$$

where Z_ℓ and Z_j are the internal degeneracies of each kind of defect.

Let us apply these considerations to our particular case of vacancies and interstitials on the A and B sublattices. Since there are four defects and one relation (6.69) between them, there are only three independent relations of the type (6.75). To solve this system, we will use the following notations:

$$\frac{n_a^v\, n_b^v}{N^2} = K \quad ; \quad K = \exp\left(-\frac{G_a^v + G_b^v}{kT}\right)$$

$$\frac{n_a^v\, n_a^i}{N^2} = K_a \quad ; \quad K_a = z \exp\left(-\frac{G_a^v + G_a^i}{kT}\right)$$

$$\frac{n_b^v n_b^i}{N^2} = K_b \quad ; \quad K_b = z \exp\left(-\frac{G_b^v + G_b^i}{kT}\right), \tag{6.76}$$

which all correspond to (6.72), i.e., to the + sign in (6.75). The pairs considered in (6.76) are the Schottky pair and the Frenkel pairs on each sublattice. We have assumed that there are z equivalent interstitial sites per normal lattice site. Using (6.76) we can express each concentration n_b^v/N, n_a^i/N, and n_b^i/N in term of n_a^v/N and inject these values into the stoichiometric condition (6.69). We then calculate n_a^v/N and finally the other three concentrations. We obtain

$$\frac{n_a^v}{N} = \sqrt{KK'} \quad , \quad \frac{n_b^v}{N} = \sqrt{\frac{K}{K'}}$$

$$\frac{n_a^i}{N} = \frac{K_a}{\sqrt{KK'}} \quad , \quad \frac{n_b^i}{N} = K_b \sqrt{\frac{K}{K'}} \tag{6.77}$$

with

$$K' = \frac{K+K_a}{K+K_b} \quad . \tag{6.78}$$

The situation discussed above, in which the compound has a composition corresponding to the stoichiometric ratio, is actually an ideal limit. In practice, the compound contains a small excess or deficit of one of the components. The variation of compound composition involves the transfer of atoms between the material and another phase, usually a vapor or a liquid phase. It is obtained from the reaction involving crystal-vapor or crystal-liquid equilibrium. In case of crystal-vapor equilibrium, the relations between the partial pressures of the components and the concentrations of these components in the compound replace (6.69), which expresses the stoichiometric condition[3].

---

[3] For considerations on phase diagrams see [6.1,11].

6.3 On the Nature of the Defects Present at Thermal Equilibrium

What is the nature of the defects present at thermal equilibrium? They should of course be intrinsic defects, i.e., vacancies, interstitials, vacancy-interstitial pairs, or complexes of vacancies or interstitials. Since the association of several vacancies or interstitials must have a free energy larger than the free energy of one of its constituent, the concentration of these complexes should be negligible, compared to the concentration of single vacancies, interstitials, or vacancy-interstitial pairs. Consequently, we shall not consider them.

According to the relations derived in Sect.6.2, the concentrations of vacancies (v) and of interstitials (i) are respectively

$$n_v = N \exp\left(\frac{S_F^v}{k}\right) \exp\left(-\frac{H_F^v}{kT}\right) \tag{6.79}$$

$$n_i = Z N \exp\left(\frac{S_F^i}{k}\right) \exp\left(-\frac{H_F^i}{kT}\right) . \tag{6.80}$$

In these relations, Z is the number of equivalent interstitial positions per normal site of the crystal. The free energy is written as the sum of the enthalpy term H_F and entropy term $-TS_F$. It is assumed that the defects are in small concentration and do not interact. When the three types of defects (vacancies, interstitials, and vacancy-interstitial pairs) coexist, we have according to (6.39) the following relation between n_{vi}, n_v, and n_i:

$$n_{vi} = \frac{n_v n_i}{N} \exp\left(\frac{\Delta G}{kT}\right) , \tag{6.81}$$

or

$$n_{vi} = \frac{n_v n_i}{N} \exp\left(-\frac{\Delta S}{k}\right) \exp\left(\frac{\Delta H}{kT}\right) , \tag{6.82}$$

where $\Delta G = \Delta H - T\Delta S$ is the binding free energy between a vacancy and an interstitial.

To compare n_i, n_v, and n_{vi}, it would be necessary to know at least order of magnitudes for the various enthalpies and entropies involved. Since the number Z of possible interstitials sites per normal crystal site is of the order of unity, we take Z = 1. Assume first that all the entropy terms are of the same order of magnitude so that all the preexponential factors in the expression giving the concentration are practically equal. Now, consider the case where $H_F^i < H_F^v$, as is predicted by various theoretical results (the

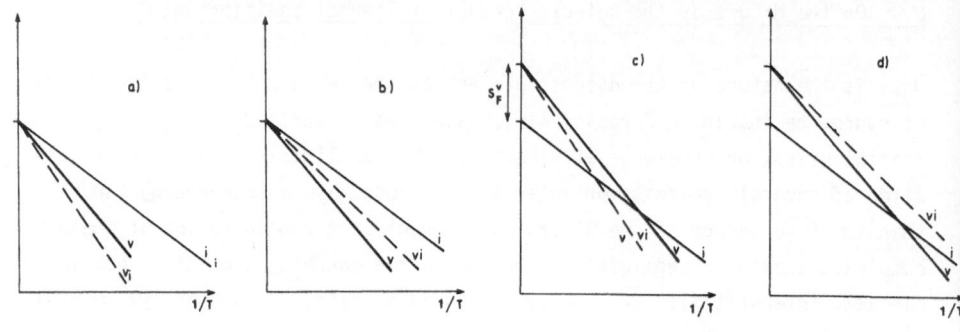

<u>Fig.6.4a-c.</u> Variation of the concentration of vacancies (v), interstitials (i), and vacancy-interstitial pairs (vi) versus temperature in an Arrhenius plot. It is assumed that $H_F^i < H_F^v$. Cases (a) and (c) correspond to $\Delta H < H_F^v$ and cases (b) and (d) to $\Delta H > H_F^v$. In cases (a) and (b) all the entropy terms are assumed to be equal while in cases (c) and (d) the entropy associated with the vacancy is assumed to be larger than the entropy of the interstitial

other case would lead to the same conclusions once the i and v indexes were inverted). Then the defect which dominates is the interstitial followed by the vacancy when $\Delta H < H_F^v$ (Fig.6.4a) and by the vacancy interstitial pair when $\Delta H > H_F^v$ (Fig.6.4b). However, these conclusions can be completely modified if the entropy terms are drastically different. Consider, for instance, the case where the entropy of the vacancy is very large compared to the entropy of the interstitial. Then, using the fact that $S_F^{vi} \simeq S_F^v + S_F^i$, i.e., that ΔS is negligible for $\Delta H < H_F^v$, the vacancy dominates in the high-temperature range and the interstitial in the low-temperature range (Fig.6.4c); for $\Delta H > H_F^v$ the vacancy interstitial pairs dominate (Fig.6.4d).

In practice, Fig.6.3 allows us to make an evaluation of the higher limit of ΔH. At high temperature (where the concentration of thermally generated defects is not negligible) and for close-pairs (where an order of magnitude for the closest vacancy-interstitial pair distance is r ~ 3 Å), we have $\Delta H \sim 0.1$ eV in silicon for the elastic interaction and ~0.3 eV for the Coulomb interaction. It is therefore reasonable to think, considering the values of the enthalpies observed experimentally (see next section), that we are in the case $\Delta H < H_F^v, H_F^i$. Consequently, the real situation should be depicted by Fig.6.4a or Fig.6.4c. Moreover, because a very large entropy term seems to be associated with the vacancy (see Chap.5), the real situation might be those of Fig.6.4c.

6.4 Experimental Determination of Enthalpies

The way to obtain information on the intrinsic defects present at high temperature is to use a physical property sensitive to their presence. By comparison between the value of this property at a given (high) temperature and
the value extrapolated from low temperature (where the concentration of defects is negligible), it is, in principle, possible to deduce the contribution
of the defects. Unfortunately, in semiconductors the concentration of the
defects thermally generated is rather small, even for temperatures close to
the melting temperature. Indeed the entropy term associated with defect formation is found experimentally to vary, from 1 to 10 K, while the enthalpy is
of the order of 2 to 5 eV. Typically, these values result in a defect concentration at 1500 K not larger than 10^{16} cm^{-3}. Consequently the defects
cannot be studied at the temperature where they are in equilibrium: their
concentration is too small to induce measurable changes in the lattice parameter or in the crystal expansion. The change in carrier concentration they
could induce at high temperature (by trapping free carriers) is very small
compared to the free carrier concentration at high temperature ($\sim 10^{20}$ cm^{-3}
at 1500 K in silicon) and therefore cannot be measured.

The only way to obtain information on these defects is to quench them
from high-temperature and study, in a low temperature range, a physical
property sensitive to a change in defect concentration. One such property
is the free carrier concentration; typically, at room temperature, the free
carrier concentration, which can be varied (by doping) in a very large range
(10^{11} to 10^{18} cm^{-3}), can be chosen in such a way that the defects introduced
by quenching induce a measurable change. The material is quenched from T to
the temperature of measurement T_0. The free carrier concentration (measured
by Hall effect), which was initially C_0 at T_0, is then reduced to C and the
concentration of defects introduced by the quenching is

$$n = C_0 - C \; . \tag{6.83}$$

Then, according to (6.15), the variation of $\ln(n)$ versus T^{-1} should be linear,
the slope R providing the formation enthalpy and the extrapolation of T^{-1}
to zero the entropy term. Such experiments have been performed, resulting in
R = 2.5 eV for silicon and R = 2.1 - 1.7 eV for germanium (see Table 6.5).

Such analysis assumes a) that all defects are electrically active since
the introduction of neutral defects, which do not trap free carriers, does
not modify the carrier concentration and b) that no defect is lost during

Table 6.5. Activation energy R[eV] associated with the defects introduced by quenching in silicon and germanium

	R [eV]	Ref.
Si	2.5	[6.12]
Ge	1.2	[6.13]
	1.9	[6.14]
	2	[6.15]
	2	[6.17]
	1.2 - 1.3	[6.18]
	2 - 1.5	[6.19]
	2	[6.20]
	1.6 - 2.5	[6.21]

quenching. If the quantity of the lost defects is proportional to the defects present at T, then the enthalpy R observed remains valid but not the entropy. Quenching experiments are difficult to perform in practice. Care must be taken that fast diffusing impurities (such as copper in germanium) do not contaminate the sample since these impurities are deep centers that also trap free carriers.

The authors in Table 6.5 have assumed that the thermally generated defects are vacancies and they ascribed the associated activation energy R to H_F^V. But as we discussed in the preceding section, vacancies as well as interstitials can be the dominant defects, depending on the temperature range at which the thermal equilibrium is studied. The temperature at which the defects have been quenched (room temperature) is such that vacancies and interstitials are mobile in silicon and germanium. Consequently, the defects observed can only be complexes, i.e., associations of vacancies or of interstitials with various impurities. The interpretation of a quenching experiment is therefore questionable: it is possible that the complex defects which are formed during quenching do not correspond to the dominant defects but to the defects which have the largest ability to be trapped by impurities forming stable complexes. For instance, interstitials can be the dominant defects, but the defects present after quenching can be vacancy-impurity complexes if the interstitial is unable to form stable complexes at room temperature. The complexes formed by quenching have not been identified although this could be possible since now there are experimental techniques which allow such identification [1.1]. In silicon, there are many vacancy-impurity complexes which are known to be stable at room temperature, but the self-interstitial has apparently not been observed, even under the form of interstitial related defects. The reason why complexes involving interstitials have never been observed is not clear, but could be the consequence that, for most of them, the associated binding energy is smaller than kT (at room temperature). Consequently, their concentration is always small compared to the concentration of vacancy complexes. Such considerations imply that R should be ascribed to H_F^V even if vacancies are not the dominant defects. Therefore, we shall now consider if this conclusion is compatible with other experimental results. The results which provide information on the formation and migration enthalpies of the vacancy are self-diffusion experiments and annealing experiments. The kinetics of annealing following electron irradiation at 4 K, which creates isolated vacancies, allow the determination of the migration enthalpy H_m^V of this vacancy. Self-diffusion, because it occurs through a vacancy mechanism, is characterized by an activation energy Q, which is the sum (see Chap.7)

$$Q = H_F^V + H_m^V \quad . \tag{6.84}$$

The values of these quantities are given for silicon and germanium in Table 6.6. As shown in this table, the values of Q and H_m^V, from which H_F^V is deduced using (6.84), are not in agreement with the fact that $R = H_F^V$. The only way to get a possible agreement between quenching, annealing and self-diffusion data is to assume that the (electrically active) defects which are observed after quenching are a) interstitial complexes, in which case $R = H_F^i$ or b) a sum of interstitial and vacancy complexes, in which case R is a combination of H_F^V and H_F^i. The identification of the defects obtained by quenching should therefore be made before any reasonable conclusion can be drawn from experiments.

Table 6.6. Activation energies [eV] obtained from quenching (R), self-diffusion (Q) and vacancy-annealing (H_M^V) experiments and formation enthalpy (H_F^V) of the vacancy deduced from (6.84). The references for R are given in Table 6.5 and for Q in Chap.7

	R	Q	H_M^V	H_F^V
Ge	1.7 - 2.1	3	0.3 [6.21]	2.7
Si	2.5	4.7 - 5.1	0.2 - 0.33 [6.22]	4.4 - 4.9

6.5 The Statistical Distribution of Donor-Acceptor Pairs

We intend to derive here the distribution of donor-acceptor pairs as a function of the donor-acceptor distance R [6.23]. This distribution is characteristic of the equilibrium at the temperature T_0 where the sample is prepared and where at least one element of the pair is mobile. For simplicity we shall consider R as a continuous variable which is enough in most cases. The extension to a discrete set of values R_i is simple.

The central quantity to be calculated is the pair distribution function G(R). For this, we consider a fixed member of the pair, say the acceptor, assuming that at T_0 only the donor is mobile. The probability of finding a donor as neighbor between R and R + dR is defined as G(R) dR. At low concentrations, where other possibilities are negligible, this quantity is proportional:

   a) to the probability of finding one donor between R and R + dR,
   b) to the probability of finding no other donor at a shorter distance,
   i.e., within the sphere of radius R.

The first probability can be obtained from the theory discussed in Sect. 6.2.5. There, the concentration of pairs at a distance R was shown to be proportional to the concentration of each species and to exp(ΔG/kT), where ΔG is the binding free energy of the pair. Here ΔG is dominated by the electrostatic interaction-$e^2/\varepsilon R$ when, at T_0 all pairs are ionized. The probability of finding a donor between R and R + dR is then given by

$$4\pi R^2 N dR \, \exp\left(\frac{e^2}{\varepsilon R k T}\right) \tag{6.85}$$

where N is the concentration of donors. The second probability is equal to one minus the probability of finding a donor within the sphere of radius R. In the case of no Coulomb interaction we should obtain

$$1 - \int_{R_0}^{R} G(R') \, dR' \quad , \tag{6.86}$$

where R_0 is the distance of closest approach of the donor. However, because of the Coulomb interaction, the simultaneous probability of finding a donor between R and R + dR and another one within the sphere should be reduced by the factor exp($-e^2/\varepsilon R k T$), since the Coulomb repulsion of the two donors is, on the average, $e^2/\varepsilon R$. Taking this into account leads to the corrected probability of finding no donor within R:

$$1 - \int_{R_0}^{R} G(R') \exp\left(- \frac{e^2}{\varepsilon R' kT}\right) dR' \quad . \tag{6.87}$$

We then obtain for $G(R)$

$$G(R) = 4\pi R^2 N \exp\left(\frac{e^2}{\varepsilon R kT}\right)\left[1 - \int_{R_0}^{R} G(R') \exp\left(- \frac{e^2}{\varepsilon R' kT}\right)\right] \quad . \tag{6.88}$$

This expression can be differentiated, resulting in

$$\frac{d}{dR}\left[\frac{G(R)}{4\pi R^2 N \exp\left(\frac{e^2}{\varepsilon R kT}\right)}\right] = - \frac{G(R)}{\exp\left(\frac{e^2}{\varepsilon R kT}\right)} \quad , \tag{6.89}$$

or

$$\frac{d}{dR} \log\left[\frac{G(R)}{4\pi R^2 N \exp\left(\frac{e^2}{\varepsilon R kT}\right)}\right] = - 4\pi R^2 N \quad , \tag{6.90}$$

leading, after integration, to

$$G(R) = 4\pi R^2 NC \exp\left(\frac{e^2}{\varepsilon R kT}\right) \exp\left(- \frac{4\pi R^3 N}{3}\right) \quad , \tag{6.91}$$

where C is an integration constant to be determined by normalization. The logarithmic derivative of G gives

$$\frac{dG/dR}{G} = \frac{2}{R} - \frac{e^2}{\varepsilon kT} \frac{1}{R^2} - 4\pi N R^2 \quad . \tag{6.92}$$

At short distances the first two terms dominate and there is a minimum approximately given by

$$R_c \simeq \frac{e^2}{\varepsilon kT} \quad . \tag{6.93}$$

At larger distances the first and third terms dominate and there is a maximum at

$$R_m \simeq \left(\frac{1}{2\pi N}\right)^{1/3} \quad . \tag{6.94}$$

The shape of $G(R)$ is shown in Fig.6.5. We can consider two distinct kinds of pairs:

  a) the pairs for which $R/R_c < 1$, corresponding to

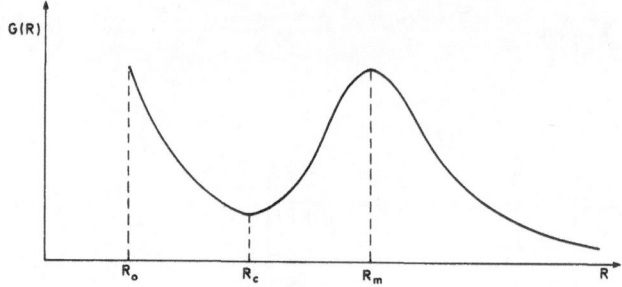

<u>Fig.6.5.</u> Pair distribution function (i.e., number of pairs of defects sep-
arated by a distance R) in case of Coulomb attraction

$$\frac{e^2}{\varepsilon R} > kT \quad . \tag{6.95}$$

These are truly associated pairs for which the binding energy is larger than
kT.

b) the pairs for which $R > R_c$. They correspond in fact to unassociated
pairs; their statistical distribution is independent of any binding energy. This becomes evident if the Coulomb interaction is reduced to zero,
taking the mathematical limit $e \to 0$ in (6.91). In this limit we would
find

$$G(R) = 4\pi R^2 N \, \exp\!\left(-\frac{4}{3}\,\pi R^3 N\right) \quad , \tag{6.96}$$

which is the normalized pair distribution function for a random distribution of donors.

When $R_m > R_c$, i.e., at low concentrations, the two kinds of pairs are
distinct and G(R) practically reduces to the sum of each separate distribution.

7. Defect Migration and Diffusion

There are various mechanisms that allow a defect to move through a lattice. These mechanisms belong to two classes, depending on whether the defect is substitutional or interstitial. An interstitial defect migrates by jumping from its original interstitial site to a neighboring equivalent one. This is illustrated in Fig.7.1 for a two-dimensional lattice, but in a real crystal the jumps occur, of course, in a three-dimensional lattice. The interstitial can also exchange with a lattice atom which in turn is displaced into a new interstitial site; this mechanism is called the interstitialcy or dumbell mechanism.

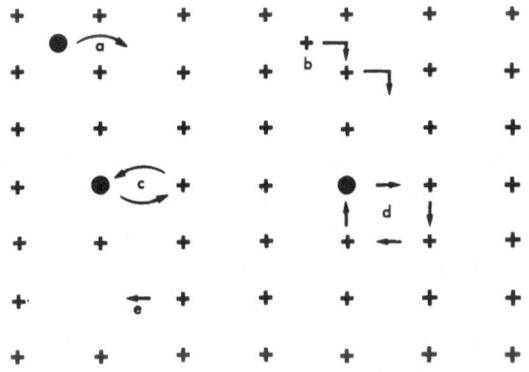

Fig.7.1a-e. Schematic representation of diffusion mechanisms: interstitial mechanism (a), interstitialcy mechanism (b), exchange mechanism (c), ring mechanism (d) and vacancy mechanism (e)

A substitutional defect migrates by jumping from its original site to a neighboring substitutional site. This can, in principle, occur through a direct exchange or through an indirect exchange (the ring mechanism). But because the indirect exchange necessitates the simultaneous jumps of two or more atoms, it is less probable than the vacancy mechanism. In the vacancy mechanism the defect jumps into a neighboring site when this site becomes empty, due to the thermal generation of vacancies.

An interstitial mechanism requires for the migration only a jump over a barrier G_m and, as is shown later, the jump probability is proportional to $\exp(-G_m/kT)$. In a vacancy mechanism the total jump probability is the product of the probability to jump over the barrier times the probability [given by (6.15)] of finding a vacancy on a neighboring site. It is therefore proportional to

$$\exp\left(- \frac{G_m}{kT}\right) \cdot \exp\left(- \frac{G_F^V}{kT}\right) \quad . \tag{7.1}$$

The identification of a particular migration mechanism is a difficult problem. Because an individual atom inside a crystal cannot be observed directly as it migrates, the only information we can obtain on its migration is necessarily indirect.

In this chapter we shall discuss the evaluation of migration enthalpies and of jump probabilities, the formation enthalpies having been considered in Chap.6. We shall also examine the main factors which play a role in migration processes, in particular, charge-state effects, and we shall describe the consequences they have on self-diffusion and impurity diffusion.

7.1 Jump Probability and Migration Energy

The migration of an atom occurs when it moves from one of its stable positions Q to a neighboring equivalent one R. In order to calculate the rate of this process, the knowledge of the energy as a function of the atomic positions is required. This can be done simply only in the framework of the Born-Oppenheimer approximation where the electron and nuclei motions are decoupled. Then, the energy of the electronic system is calculated for any given set of nuclear coordinates and it is this energy which plays the role of the potential energy for nuclear motion.

Let us call x the coordinate along the axis which joins the two equivalent sites Q and R, and x_i the set of all other atomic coordinates (for a system of N atoms, i will then vary from 1 to 3N-1). The potential energy for nuclear motion $V(x,x_i)$ will be some function of this set of coordinates. For a fixed value of x we can minimize V with respect to all x_i. When such a minimum exists (this is the usual assumption), we label it $V(x,x_{im})$ where the x_{im} are functions of x. The corresponding variation of $V(x,x_{im})$ is drawn in Fig.7.2. The quantity ΔV is the usual potential barrier for migration. The problem is to determine the rate K at which an atom initially at $Q(x = 0)$ migrates to the position $R(x = a)$. Intuitively, we can anticipate that this rate will be of the form

$$K = \nu \exp(- \Delta V/kT) \quad , \tag{7.2}$$

$\exp(- \Delta V/kT)$ being the probability of excitation over the barrier and ν some effective frequency.

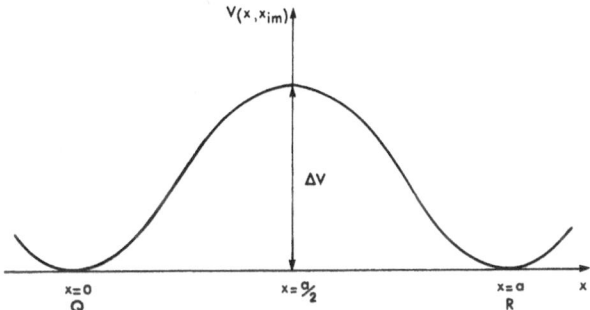

Fig.7.2. Sketch of the potential barrier along the migration path

Two distinct methods are used to determine this activation law for migration, the rate theory and the dynamical theory.

7.1.1 The Rate Theory

The initial versions of this theory are directly derived from the theories of the rate of chemical reactions. They assume that between the initial and final states Q and R exists an activated state (S at $x = a/2$ on Fig.7.2) and that equilibrium statistical mechanics apply to all these states. But, instead of this conventional approach, we follow a more recent derivation [7.1] which presents two advantages: a) it gives unambiguous definitions of ΔV and ν and b) it provides results completely equivalent to the dynamical approach after some further simplifications are made.

To describe the method, let us label v and v_i the velocities corresponding to the coordinates x and x_i. The following calculation is completely classical (quantum effects correspond to tunneling through the barrier and lead usually to small corrections [7.2]). The energy E is then

$$E = mv^2/2 + \sum_i mv_i^2/2 + V(x,x_i) \quad . \tag{7.3}$$

The rate K, at which transitions from Q to R occur, is given by the flux of particles crossing a/2 from the left to the right, i.e., (with $v > 0$)

$$K = \frac{\displaystyle\int_0^\infty v\, dv \int \ldots \int \prod_i dv_i dx_i W(x = a/2, v, x_i, v_i)}{\displaystyle\int_{-\infty}^{+\infty} dv \int \ldots \int dx \prod_i dv_i dx_i W(x, v, x_i, v_i)} \quad , \tag{7.4}$$

where $\prod_i dv_i\, dx_i$ is the product over i of all the different elements and $W(x,v,x_i,v_i)$ the probability of finding the system in the state corresponding to this set of coordinates. In Boltzmann statistics W is simply proportional to

$$\exp\left(-\frac{mv^2/2 + \sum_i m_i v_i^2/2 + V(x,x_i)}{kT}\right) \quad . \tag{7.5}$$

Integration over the v_i cancels between the numerator and the denominator in (7.4). Therefore, after integration over v, K comes out to be

$$K = \frac{1}{\sqrt{2\pi m}} \frac{\displaystyle\int \ldots \int \exp\left(-\frac{V(x = a/2, x_i)}{kT}\right) \prod_i dx_i}{\displaystyle\int \ldots \int \exp\left(-\frac{V(x,x_i)}{kT}\right) dx \prod_i dx_i} \quad . \tag{7.6}$$

This expression is general, but not very useful. To go further we exploit the fact that kT is usually small enough so that the main contributions to the integrals come from regions near the potential minimum. For the denominator we expand V with respect to one of its absolute minima Q

$$V(x,x_i) \simeq V(Q) + \frac{1}{2} \sum_{\alpha,\beta} \left(\frac{\partial^2 V}{\partial u_\alpha \partial u_\beta}\right)_0 u_\alpha u_\beta \quad , \tag{7.7}$$

where the u_α and u_β are the coordinates of the atomic displacements from Q. As we have seen in Chap.5 $(\partial^2 V/\partial u_\alpha \partial u_\beta)_0$ defines the matrix of force constants A which is symmetrical and can be diagonalized. Let us call $A(\lambda)$ and $u(\lambda)$ the eigenvalues and eigenvectors of A. The denominator D in (7.6) can be integrated with respect to these new coordinates $u(\lambda)$, giving

$$D = \exp\left[-\frac{V(Q)}{kT}\right] \int_{-\infty}^{+\infty} \cdots \int_{-\infty}^{+\infty} \exp\left[-1/2 \cdot \sum_{\lambda} \frac{A(\lambda)}{kT} u^2(\lambda)\right] \prod_{\lambda} du(\lambda) \quad , \tag{7.8}$$

that is

$$D = \exp\left[-\frac{V(Q)}{kT}\right] \prod_{\lambda=1}^{3N} \left[\frac{2\pi kT}{A(\lambda)}\right]^{\frac{1}{2}} \quad . \tag{7.9}$$

The same arguments can be used to evaluate the numerator N in (7.6) when V(S) is a saddle point. However x is now fixed at a/2, and the integration is now performed over 3N-1 coordinates. This leads to

$$N = \exp\left[-\frac{V(S)}{kT}\right] \prod_{k=1}^{3N-1} \left[\frac{2\pi kT}{B(k)}\right]^{\frac{1}{2}} \quad , \tag{7.10}$$

where B is the matrix of force constants at the saddle point (with 3N-1 degrees of freedom). Then, the rate K turns out to be

$$K = \left[\frac{1}{m} \frac{\prod\limits_{\lambda=1}^{3N} A(\lambda)}{\prod\limits_{k=1}^{3N-1} B(k)}\right]^{\frac{1}{2}} \exp\left(-\frac{\Delta V}{kT}\right) \quad . \tag{7.11}$$

This expression defines unambiguously the preexponential factor. The ratio $\Pi A/\Pi B$ has the dimension of an effective force constant C. The rate K is of the expected form $\nu \exp(-\Delta V/kT)$ with ν expressed as $\sqrt{C/m}$.

The expression (7.11) leads to the important conclusion that the frequency ν is proportional to $1/\sqrt{m}$, i.e., to the inverse square root of the mass of the diffusing atom. This gives a direct test of the validity of the theory. The preexponential factor does not directly involve a vibrational entropy, but is in some respects equivalent to it. Finally, all the derivation was made with the implicit assumption of constant volume V. As diffusion occurs at constant pressure p a term pV should be added to the energy, leading for the barrier height to a difference in enthalpy H_m instead of internal energy ΔV, so that the jump probability becomes

$$K = \nu \exp\left(-\frac{H_m}{kT}\right) \quad . \tag{7.12}$$

Expression (7.12) does not involve a free energy $H_m - TS_m$. The equivalent of the entropy part is included in the definition of the frequency ν given by (7.11). To realize this we write ν under the form

$$\nu = \nu_0 \left[\frac{\displaystyle\prod_{\lambda=1}^{3N} A(\lambda)}{B_0 \displaystyle\prod_{k=1}^{3N-1} B(k)} \right]^{\frac{1}{2}} , \qquad (7.13)$$

where ν_0, equal to $\sqrt{B_0/m}$, is the vibration frequency of the migrating atom near the point Q of Fig.7.2, all other atoms being fixed. Since it is always possible to express (7.13) under the form

$$\nu = \nu_0 \exp\left(\frac{S_m}{k}\right) , \qquad (7.14)$$

we have

$$\frac{S_m}{k} = \frac{1}{2} \left\{ \sum_{\lambda=1}^{3N} \log A(\lambda) - \left[\log B_0 + \sum_{k=1}^{3N-1} \log B(k) \right] \right\} , \qquad (7.15)$$

which closely resembles the usual definition of a vibrational entropy (see Chap.5). In this way we recover for ν an expression of the form $\nu_0 \exp(-G_m/kT)$ where G_m is equal to $H_m - TS_m$. However, the partitioning of (7.13) is not uniqu and G_m is not an unambiguously defined free energy.

Within the rate theory, the defect enthalpy H_m for migration is the difference between the defect formation enthalpies in the saddle point (S) and in the equilibrium position. The theoretical determination of H_m is therefore obtained through the estimation of formation energies. As discussed in Chap.6, the estimations of formation energies made up to now are only qualitative and this is particularly true for migration energies.

It is therefore useful to estimate H_m from very crude approximations. The simplest of them consists in the replacement of the actual potential wells by their parabolic approximation (Fig.7.3). The migration enthalpy H_{0m}, is given by their intersection, and if the parabola at Q is given by $kx^2/2$, we can write

$$H_{0m} = \frac{ka^2}{8} . \qquad (7.16)$$

Figure (7.3) shows clearly that H_{0m} is a crude overestimation. It might be better to approximate $\Delta H(x)$ by some periodic function. Taking a cosine function,

$$\Delta H(x) = \frac{H_m}{2} \left[1 - \cos\left(2\pi \frac{x}{a}\right) \right] , \qquad (7.17)$$

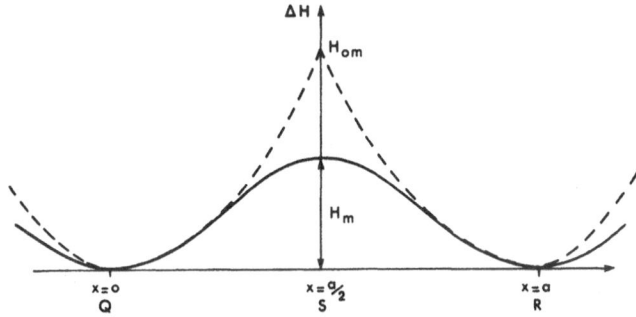

Fig.7.3. The parabolic (---) and periodic approximations to the potential barrier (——)

H_m is determined by the condition that near Q, $\Delta H(x)$ reduces to the correct parabolic form $kx^2/2$, i.e.,

$$H_m = \frac{ka^2}{2\pi^2} \; , \tag{7.18}$$

H_m is about 2.5 times smaller than H_{Om}.

The characteristic constant k of the harmonic oscillator can be related to the vibrational frequencies by the techniques developed in Chap.5. Let us apply them to Group IV semiconductors, using the phonon Hamiltonian given in Sect.5.1.4. The main part of this Hamiltonian is $1/2 \cdot k_{rr} \sum_{ij} (dr_{ij})^2$, where k_{rr} is the radial force constant between nearest neighbors and dr_{ij}, the increase in distance between neighbors i and j. For the perfect crystal we have shown that the maximum frequency ω_M is related to k_{rr} by

$$\omega_M^2 = \frac{8}{3} \frac{k_{rr}}{m_0} \; , \tag{7.19}$$

m_0 being the atomic mass in the perfect crystal. In the same model, (5.32) allows k to be obtained under the form

$$k = k_{rr} \sum_j x_{0j}^2 \; , \tag{7.20}$$

where X_{0j} is the component along the migrating path $\vec{OX}$ of the unit vector joining the atom at position Q to its nearest neighbor j. In a tetrahedral crystal for a migrating atom next to a vacancy, there are only three neighbors j, with X_{0j} equal to -1/3. Thus,

$$k = k_{rr}/3 \; , \tag{7.21}$$

that is,

$$k = \frac{m_0 \omega_M^2}{8} \quad . \tag{7.22}$$

Taking the cosine law leading to the (7.18) for H_m, we obtain H_m in terms of ω_M

$$H_m = \frac{m_0 \omega_M^2 a^2}{16\pi^2} \quad . \tag{7.23}$$

7.1.2 The Dynamical Theory

A second useful approach to the description of atomic migration is known as the dynamical theory [7.1,3]. The viewpoint is somewhat different from that of the rate theory. It consists of treating the displacements causing migration as a superposition of phonons in an harmonic crystal. The jump occurs when the individual displacements of the defect induced by phonons add to produce a large local displacement. For instance, a defect jumps when it has a sufficiently large amplitude of motion in the direction of the jump and when, in the same time, the neighboring atoms at the saddle point move sufficiently to reduce closed-shell repulsion between them and the defect. We do not reproduce here the details of the whole theory based on a normal mode analysis of the system. Fluctuations in velocities and mode amplitudes follow Gaussian distributions. Again the rate at which the migrating atom reaches the point $x = a/2$ with a positive velocity is evaluated. It has been shown [7.4] that the complete result is equivalent to (7.6) when the potential occuring in the numerator as well as the denominator is expanded to second order in the atomic displacements about the position Q.

Useful information can be obtained by using simplifying arguments. This was done specifically to derive a relation between H_m and the Debye temperature. As this is typical of the dynamical point of view, we reproduce it briefly. For instance, the probability for the migrating atom to be displaced at x from its original position $Q(x=0)$ is given by

$$P(x) = \frac{1}{\sqrt{2\pi} \langle x^2 \rangle} \exp\left(-\frac{1}{2} \frac{x^2}{\langle x^2 \rangle}\right) \quad , \tag{7.24}$$

where $\langle x^2 \rangle$ is the mean square displacement, i.e., some thermal average over x^2. This mean displacement is related to the frequency spectrum of the crystal and then in the Debye approximation, to the Debye frequency ω_D (or Debye temperature θ_D). It has been shown [7.4] that $<x^2>$ is given by

$$<x^2> = \frac{3kT}{m_0\omega_D^2} \quad .$$ (7.25)

This value for $<x^2>$ injected into (7.24) gives an argument of the exponential equal to

$$\frac{1}{6}\left(\frac{k}{\hbar}\right)^2 \frac{m_0\theta_D^2 x^2}{kT} \quad .$$ (7.26)

As the jump takes place at $x = a/2$, the activation energy becomes

$$H_m = \frac{m_0\omega_D^2 a^2}{24} \quad .$$ (7.27)

This expression has to be corrected for displacements of the neighboring atoms caused by the jump. This leads to a corrective factor in H_m close to unity.

Expression (7.27) has been tested for a number of metals. It is found [7.4] that the ratio $H_m/(m_0\omega_D^2 a^2)$ is practically constant and equal to 10^{-2}, i.e., four times smaller than the theoretical prediction. One reason for this presumably is that (7.27) corresponds to a value of H_m close to H_{0m} (Fig. 7.3), since it is deduced from the harmonic approximation about the stable positions. Corrections to (7.27) have been brought by the "ballistic model" [7.5]. The arguments for this model are derived from the rate theory, but remain qualitative, the proposed formula being

$$H_m = F^2 \frac{m_0\omega_D^2 a^2}{8\pi^2} \quad .$$ (7.28)

F is an empirical factor of order unity, which corrects for structure effects. Clearly (7.28) is an ad hoc correction to (7.27), reducing it by about the correct factor.

It is interesting to compare the semi-empirical formula (7.28) with the previously derived (7.23). As shown in Table 7.1, the Debye frequency ω_D and the Raman frequency ω_M are nearly equal in the sequence C, Si, Ge, Sn. This means that the enthalpy of migration calculated from (7.23) is about one half of the enthalpy derived from (7.28) (this is not quite true since F is taken to be 0.9 for the diamond structure).

Table 7.1. Debye temperature θ, vibrational frequency ν_D, Raman frequency ν_M, interatomic distance a, atomic mass A and migration enthalpy H_m as obtained from (7.28) and (7.23)

	θ [K]	ν_D [×10^{12} Hz]	ν_M [×10^{12} Hz]	a [Å]	A [amu]	H_m from (7.28) [eV]	H_m from (7.23) [eV]
Diamond	2230	46	40	1.54	12	2.6	1.2
Silicon	645	13.5	15	2.35	28.1	1.18	0.9
Germanium	374	8	9	2.45	72.6	1.11	0.83

7.2 Experimental Determination of Migration Enthalpies

The above approximations seem to provide reasonable results if we compare them with the theoretical estimates obtained from the classical or quantum-mechanical treatments described in Chap.6 (see Table 7.2). However, they are not at all in agreement with the few experimental results available.

Table 7.2. Theoretical enthalpies [eV] of migration calculated for the vacancy (V) and the interstitial (I)

	Germanium	silicon	diamond	reference
H_m^V	0.95	1.06		[7.6]
	0.98	1.09		[7.7]
	0.062	0.68	1.28	[7.8]
H_m^I	0.44	0.51		[7.7]
	<0.25	<0.22		[7.9]
			0.17	[7.10]
	0.09	0.18		[7.11]

The way a migration enthalpy of a defect can be measured is the following. First, the defect must be created at a temperature at which $H_m \gg kT$ in such a way that the defect remains immobile. This can be done by quenching (Sect. 6.4.1) or by irradiation [Ref.1.1, Chap.12]. Then the temperature is raised up to a value for which defect mobility occurs. The migration enthalpy is obtained from the study of the kinetics of annealing [Ref.1.1, Chap.13]. The

annealing reflects the change in defect concentration due to defect annihilation or association on various sinks. The annealing process being thermally
activated, the associated activation energy is the migration enthalpy.

As an illustration consider the case of the vacancy in silicon. Electron
irradiation at low temperature (4 K) creates simple point defects. One of
these defects has been identified, using electron paramagnetic resonance
(EPR), as being the single vacancy [7.12,13]. When the temperature is raised,
in the range 80-100 K, the concentration of these single vacancies decreases:
the vacancies disappear because they become mobile and associate with impurities as witnessed by the formation of A centers (oxygen-vacancy complexes),
E centers (doping-impurity - vacancy complexes) and divacancies (also identified by EPR). The study of the kinetics of the disappearance of the vacancies
therefore provides the migration enthalpy of the vacancy. It is found to be
0.18 eV or 0.33 eV depending on the Fermi-level position, i.e., on the charge
state of the vacancy.

In the case of a complex defect it is necessary to verify that the annealing process is indeed due to defect mobility and not to its dissociation before the activation energy can be ascribed to the migration enthalpy of the
defect.

In Group IV semiconductors the determination of the migration enthalpy of
simple intrinsic defects, vacancies, and self-interstitials, necessitates the
use of irradiations performed below room temperature. The free self-interstitial has not been identified in silicon or germanium, and practically no
information has been obtained on it. As to the vacancy, it has been firmly
identified only in silicon. In germanium the vacancy has not been identified,
but presumably becomes mobile around 90 K with an activation energy of the
same order of magnitude as for the vacancy in silicon [7.14]. In diamond the
vacancy associated with the so-called GR optical system anneals in several
stages up to 1200 K and consequently, the associated activation energies
cannot be ascribed to the migration enthalpy of this vacancy [7.15].

The experimental results in the case of silicon do not agree at all with
any of the theoretical estimates (Table 7.2). It is evident that these estimates should take into account charge-state effects which apparently play an
important role: a change of charge state, from V^0 to V^+, induces a change of
~50% in the migration enthalpy of the vacancy in silicon. In the next section
we develop a classical theory which illustrates in a simple qualitative way
how these charge-state effects can be included in the migration energy and
the possible effects they can induce on defect behavior during migration.

This theory, due to WEISER [7.16] was originally applied for charged interstitials, but can readily be generalized to any kind of defect.

7.3 Charge-State Effects on Defect Migration

7.3.1 Weiser's Theory

WEISER [7.16] argues that the interactions between a charged interstitial,
i.e., an ion, and the host atoms are composed of a) an attractive potential
produced by the interaction of the ion with the dipoles it induces in the
lattice, and b) a repulsive potential due to the nucleus-nucleus interactions.
The calculation of lattice polarization by the charge q of the ion is based
on a proposal by MOTT and LITTLETON [7.17] which they used to obtain the formation enthalpy of a vacancy in ionic crystals. The polarization energy

$$U_p = \frac{1}{2} q\phi \qquad (7.29)$$

is calculated assuming that the potential ϕ produced by the dipoles μ induced
on the surrounding atoms can be divided into two components: a discrete component for a few layers of atoms i (situated a distance R_i from the ion) and
a component due to the rest of the lattice at a distance larger than R_0,
treated as a continuum. Then,

$$\phi = - \sum_i \frac{\mu_i}{R_i^2} - \frac{q}{R_0}\left(1 - \frac{1}{\epsilon}\right) \ . \qquad (7.30)$$

The repulsive energy U_R, between the closed shells of the ion (of ionic
radius r_i) and a host atom (of ionic radius r_h) separated by a distance r,
is taken to be

$$U_R = \alpha \exp[(- r + r_h + r_i)/\rho] \ , \qquad (7.31)$$

where α is a scalar factor. The difficulty lies mostly in the choice of the
constant ρ. Values of ionic radii have been tabulated by PAULING [7.18].

Once the equilibrium (Q) and saddle point (S) configurations are chosen,
the migration enthalpy H_m is calculated from

$$H_m = \left(U_R^S - U_P^S\right) - \left(U_R^Q - U_P^Q\right) \ , \qquad (7.32)$$

where U_R and U_P are positive quantities, that is,

$$H_m = \Delta U_R - \Delta U_P \quad , \tag{7.33}$$

$$\text{with} \quad \Delta U_R = U_R^S - U_R^Q \tag{7.34}$$

$$\text{and} \quad \Delta U_P = U_P^S - U_P^Q \quad . \tag{7.35}$$

Calculations have been performed by WEISER for various impurities assuming a) the Q configuration is tetrahedral and S is hexagonal (Chap.1), and b) these impurities induce no lattice relaxation, nor distortion. Of course, such a method is, like the others, subject to strong criticism: the results are very sensitive to the values taken for the various parameters. However, the method presents the advantage of illustrating qualitatively the effect of the charge state q of a defect on its migration. Roughly speaking, U_P varies as q^2 (since the dipoles μ_i have an amplitude which varies linearly with q), while U_R varies exponentially with the ionic radius r_i. We can therefore write

$$\Delta U_P \simeq q^2 \Delta U_P^0 \quad , \tag{7.36}$$

and

$$\Delta U_R \simeq \exp\left(\frac{r_i}{\rho}\right) \Delta U_R^0 \quad . \tag{7.37}$$

Figure 7.4 illustrates schematically the variations of ΔU_P and ΔU_R with q. We consider a positive variation of q, i.e., a decrease in the number of the electrons on the defect, in which case the ionic radius r_i decreases. It is obvious on this figure that, according to (7.33), H_m varies with the charge state. For $q < q_0$, value of the charge for which $\Delta U_P = \Delta U_R$, we have

$$\Delta U_R > \Delta U_P \quad , \tag{7.38}$$

i.e.,

$$U_R^S - U_P^S > U_R^Q - U_P^Q \quad . \tag{7.39}$$

The formation energy of the ion is larger in the S than in the Q configuration: S is the saddle point and Q, the equilibrium configuration. For $q > q_0$, the situation is reversed:

$$\Delta U_R < \Delta U_P \quad , \tag{7.40}$$

i.e., the formation energy of the ion is larger in the Q than in the S configuration: S is now the equilibrium configuration and Q the saddle point. In conclusion, WEISER's theory predicts that the migration energy of the ion is charge-state dependent; it also predicts that the equilibrium configuration

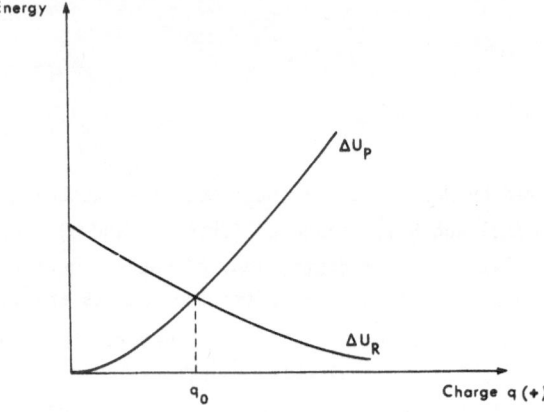

<u>Fig.7.4.</u> Schematic representation of the variation of the difference in polarization and repulsion energies between two different defect configurations, versus the charge state of the defect

of the ion can change and becomes eventually the saddle point for the migration. Of course, the charge q_0 at which this change occurs can be such that it does not exist in the semiconductor.

7.3.2 Ionization Enhanced Migration

Electrons trapped at a defect are in equilibrium with the carriers, electrons, and holes present in the conduction and valence bands. The equations which describe this equilibrium and the rate at which this equilibrium occurs will be given in Ref.[1.1] Chap.12. The charge state of a defect corresponds to the number of the carriers which remain localized on the site of the defect in addition to the electrons the defect possesses in its neutral state. The concentration of a defect in a given charge state is determined by the position of the Fermi level as compared to the position of the localized electronic state associated with the defect. It therefore depends on the temperature, the doping impurity concentration and the concentrations of other deep levels. In this section we address ourselves the question of how the presence of these free carriers, which are thermally generated or can be created by an external ionization process, can affect defect migration.

There are three types of effects possible: one is due to the presence of electrons and (or) holes in the valence and conduction bands, another is due to a change in the Fermi level position, i.e., to a change in the fraction of the defect population which is in a given charge state. The third effect is related to the fact that a particular defect, taken among the whole defect population, because it acts as a recombination center (Ref.[1.1], Chap.12) traps alternatively electrons and holes and therefore undergoes alternative charge-state changes.

a) *Electrostriction Mechanism*

The presence of electrons and holes in the conduction and valence bands reduces the bonding energy and increases the antibonding energy. Consequently, it modifies migration and formation energies. One way of getting an estimation of the energy change induced by an electron-hole pair is to calculate the dilatation d it induces and to take a first-order development of the bonding energy versus the interatomic distance r. In an elastic model, the variation dH of the enthalpy of a defect can be expressed roughly as a function of r^2 [see, for instance, (7.18)]. Then

$$\frac{dH}{H} \simeq 2 \frac{dr}{r} \quad . \tag{7.41}$$

The dilatation per electron-hole pair is such that $\sim 10^{22}$ cm^{-3} pairs are necessary to create a variation of $\sim 1\%$ in a migration or formation enthalpy [7.19]. The concentration of electron-hole pairs usually produced by light illumination, X-ray, or particle irradiation is never so high and this electrostriction mechanism should then be negligible. Only in the case of irradiation with a laser pulse of very high intensity (in the MW cm^{-2} range) will perhaps this effect be noneglibible

The dilatation also induces a change in the entropy terms. As we have shown in Chap.5, a relatively large entropy change can be the consequence of a small lattice dilatation around the defect. This entropy change can be readily calculated from the relations derived in Sect.5.4.

b) *Normal Ionization Enhanced Migration*

Consider a defect which possesses a charge q in a state B, when the Fermi level E_F is above the localized level E_T associated with this defect and a charge q' in a state S, when E_F is below E_T (i.e., q' = q + e). Let α and $\alpha' = 1 - \alpha$ be the fractions of the defect population which are in the S and B states, respectively. According to (6.33) we have

$$\frac{\alpha}{1-\alpha} = g \exp\left(\frac{E_T - E_F}{kT}\right) , \tag{7.42}$$

where g is the degeneracy factor of the level E_T. The change of charge (from q to q') induces a corresponding change in the migration enthalpy, from H_m^S to

H_m^B and consequently in the jump probability, from P_S to P_B. This situation is schematically depicted in Fig.7.5. It can be easily shown [7.19] that the total jump probability P_T is

$$P_T = \alpha P_S + (1 - \alpha)P_B \quad . \tag{7.43}$$

A well-known case where the normal ionization enhanced mechanism applies is the case of the vacancy, but there are many other examples [7.19]. In n-type silicon the vacancy, which has trapped two electrons, anneals (by migration) around 70-85 K with an activation energy of 180 meV; in p-type material the vacancy is neutral and anneals at 150-180 K with an activation energy of 330 meV.

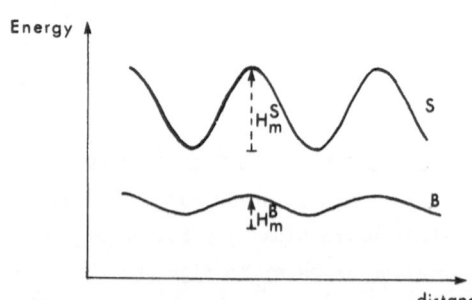

Fig.7.5. Schematic representation of the migration potential energy versus atomic distance for the normal ionization enhanced migration mechanism

c) *Athermal Ionization Enhanced Migration*

The electronic equilibrium between the defects and the bands is obtained through the thermal generation and recombination of electrons and holes. When we consider a particular member of a defect population, its average state is a temporal average obtained by a continual succession of electronic transitions between the defect state E_T and the bands (Fig.7.6). The rates of transition are given by the capture (k) and generation (g) rates of elec-

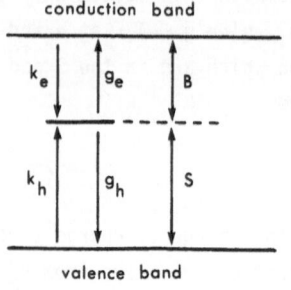

Fig.7.6. Energy level diagram showing the position of the defect level E_T with respect to the conductio and valence bands. When the Fermi level is in the energy range denoted by B (or S) the defect will be predominantly in charge state q (or q'). The capture rates for electrons and holes are denoted k_e, k_h and the generation rates g_e, g_h, respectively

trons (e) and holes (h). The expressions of these rates k_e, k_h, g_e, and g_h in terms of carrier cross section, density of states, and electronic energy level will be given in [Ref.1.1, Chap.12].

A rate of transition is the probability per unit time that a transition occurs, times the concentration of electrons in the initial state. It is therefore proportional to the concentration of empty states in the final state. These transition rates are obtained by writing the equation describing the reactions

$$S + e^- \underset{g_e}{\overset{k_e}{\rightleftharpoons}} B \quad , \tag{7.44}$$

and

$$B + h^+ \underset{g_h}{\overset{k_h}{\rightleftharpoons}} S \quad . \tag{7.45}$$

The dynamics of the rate of change of charge state is given by

$$\frac{ds}{dt} = -\frac{db}{dt} \quad , \tag{7.46}$$

and

$$\frac{ds}{dt} = (k_h + g_e)b - (k_e + g_h)s \quad , \tag{7.47}$$

s and b being the concentrations in the S and B states respectively. Using the fraction of occupancy α, the above equation reduces to

$$\frac{d\alpha}{dt} = (k_e + g_e)(1 - \alpha) - (k_e + g_h)\alpha \quad . \tag{7.48}$$

At equilibrium or in permanent regime

$$\frac{d\alpha}{dt} = 0 \tag{7.49}$$

and

$$\alpha = \frac{k_h + g_e}{k_h + g_e + k_e + g_h} \quad . \tag{7.50}$$

The occupancy of a state is obtained by a continual succession of transitions. If we consider a particular defect in the total population, it changes its charge state with an average rate $1/\tau$, which is obtained by taking the rate of change in a given state (i.e., $k_h + g_e$ for B, $k_e + g_h$ for S) times the probability of finding this state ($1 - \alpha$ for B, α for S) and summing over all possible states (two in our case). This rate is therefore given by

$$\frac{1}{\tau} = (k_h + g_e)(1 - \alpha) + (k_e + g_h)\alpha \quad . \tag{7.51}$$

Replacing α by its value given in (7.50) we obtain

$$\frac{1}{\tau} = \frac{2}{\frac{1}{k_e + g_h} + \frac{1}{k_h + g_e}} \quad . \tag{7.52}$$

Such alternative charge-state changes can lead in some cases to the athermal migration of a defect. Indeed we have seen in the preceding section that the equilibrium configuration of a defect can be charge-state dependent and that, in the case where the transition charge q_0 (introduced in Sect.7.3.1) is between the two charges q and q', then the equilibrium defect configuration becomes the saddle-point configuration and vice-versa. This situation is schematically depicted in Fig.7.7. Consider a defect initially in the state S; it is in an equilibrium position (1). When this defect passes in the B state it finds itself in a saddle-point position (2) and therefore relaxes to a new equilibrium position (3). When again the defect comes back to its original state S it must again relax from position (4) to position (1) or (5), equivalent to (1). The defect has migrated from (1) to (5). The rate of the jump is

$$\gamma = \frac{\tau^{-1}}{Z} \quad , \tag{7.53}$$

where Z is the number of equivalent sites the defect can jump into.

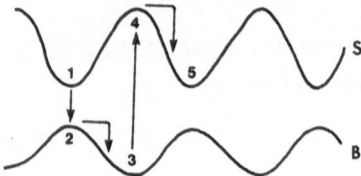

Fig.7.7. Schematic representation of the migration potential energy versus atomic distance for the athermal ionization enhanced migration mechanism showing the corresponding electronic transitions

This migration mechanism is said to be athermal because the migration does not require any thermal energy. The energy necessary to induce the jump of the defect is provided by the electronic carrier (this energy is used to modify the defect configuration).

d) *Energy Release Mechanism*

The energy ΔE brought by the captured carrier can also be released on the defect site under the form of phonons. The emitted phonons can enhance the jump probability by an factor $\exp(\Delta E/kT)$, since the expression of K (7.12) becomes

$$K = \nu \exp - \frac{H_m - \Delta E}{kT} \quad . \tag{7.54}$$

As we shall see in [Ref.1.1, Chap.12], the way phonons are emitted in such phonon-aided carrier capture process is not completely clear. There are two types of processes: the cascade capture [7.20] and the multiphonon emission [7.21]. Cascade capture presupposes a defect with many excited levels, the separation between which is smaller than possible phonon energies. The carrier capture occurs via a cascade of one-phonon-assisted transitions between two adjacent levels. In the multiphonon emission capture, the phonons are all emitted in one transition. The multiphonon emission process must be discussed in terms of the so-called configuration coordinate diagram, which will be described in [Ref.1.1, Chaps.9, 10 and 12].

7.4 Diffusion

7.4.1 Fick's Law

Defects wander through the lattice as a result of their thermal agitation.[1] The simplest way to understand the meaning of the diffusion coefficient is to consider a jump process between two adjacent (100) planes of a simple cubic lattice. Defects can jump with a probability K per unit time either to the right or the left, each jump being of distance a. Consider two adjacent lattice planes 1 and 2 containing respectively n_1 and n_2 defects per unit surface. The number of defects per unit surface jumping in a time dt from plane 1 to plane 2 is $N_1 = n_1 Kdt$. The jump probabilities to the left or to the right being equal, the net flux of defects from plane 2 to plane 1 is

$$J_{21} = \frac{N_1 - N_2}{dt} = (n_1 - n_2)K \quad . \tag{7.55}$$

Assuming the number of defect changes slowly with the distance X, we can replace this equation by the following differential equation

$$J = - Ka \frac{\partial n}{\partial x} \quad . \tag{7.56}$$

The number of defect (per unit area) is related to the concentration C (per unit volume) by

---

[1] For general treatments of diffusion see [7.22].

$$C = n/a \quad , \tag{7.57}$$

and (7.56) becomes

$$J = - a^2 K \frac{\partial c}{\partial x} \quad . \tag{7.58}$$

This equation is known as Fick's law, given in vectorial form by

$$\underline{J} = - D \underline{\nabla} c \quad , \tag{7.59}$$

where

$$D = a^2 K \tag{7.60}$$

is called the diffusion coefficient. The expression (7.60) is valid for the diffusion of substitutional atoms in all cubic structures.

Considering the expression of K (7.11) derived in the first section, this diffusion coefficient is thermally activated:

$$D = D_0 \exp(-Q/kT) \quad . \tag{7.61}$$

For a defect which diffuses independently (by an interstitial mechanism), (7.12) shows that Q is the enthalpy for migration H_m. In the case of a diffusion through a vacancy mechanism, Q is the sum of H_m and of the formation enthalpy H_F^V of the vacancy [see (7.1)]. In semiconductors, diffusion studies are a powerful tool for obtaining information on entropies and enthalpies. The Arrhenius law, i.e., the linearity of $\ln(D)$ versus T^{-1}, is often verified, from which the preexponential term D_0 (value of D extrapolated to $T^{-1} = 0$) and the activation energy Q (slope of the straight line) can be determined. The problem is that the enthalpies and entropies, once determined, must be attributed to a specific mechanism. As we shall see now there are problems concerning the attribution of a given mechanism to a specific diffusion[2].

7.4.2 Experimental Determination of a Diffusion Coefficient

In order to obtain a diffusion coefficient corresponding to a given diffusion time at a given temperature, the experimental diffusion profile is fitted to a theoretical one obtained from the solution of Fick's law using proper boundary conditions. Actually, because of oxide formation at the surface, it is often not obvious to define the correct boundary conditions. There are two

---

[2] For reviews on diffusion in elemental and compound semiconductors see [7.23].

ways of obtaining the experimental defect profile, i.e., the concentration
of the diffused defect versus depth [7.24]. The first one is destructive:
it consists in the removal of successive layers and in the measurement of
the remaining impurity concentration or of the removed impurity concentration.
The second one is nondestructive.

The destructive method involving the measurement of the concentration of
the removed impurities is based on the use of radioactive isotopes ("labelled"
atoms). The methods involving the measurement of the concentration of the remaining atoms use electrical (Hall effect, conductivity), optical (absorption)
or paramagnetic properties specific to semiconductors. Such methods lack
the universality of the active isotope methods (the impurity must be electrically active, optically active, or exhibit paramagnetic properties),
but are usually more convenient to apply. Self-diffusion can therefore only
be studied using isotope tracers.

The nondestructive methods use capacitance-voltage and backscattering
techniques. The profile of electrically active defects can be obtained from
the modification they induce in the free carrier profile using the capacitance-
voltage characteristics of a junction [Ref.1.1, Chap.12]. The depth is limited
by the free carrier concentration (the larger the carrier concentration, the
smaller the depth), i.e., by the impurity concentration which has diffused or
by the defect concentration which compensate the initial free carrier concentration. Backscattering measurements are limited to diffusing elements
heavy compared to bulk atoms and present in large enough concentrations to
allow incident particles to be backscattered with a nonnegligible counting
rate.

7.4.3 Self-Diffusion

The most natural mechanisms for self-diffusion are the vacancy mechanism
and the interstitial (or interstitialcy) mechanism. In germanium it is generally accepted that there is strong evidence in favor of a vacancy mechanism.
The self-diffusion coefficient, deduced from the rate of precipitation of
supersaturated solid solutions [7.25] assuming a vacancy mechanism, is in
agreement with the self-diffusion coefficient directly measured with the help
of radioactive tracers. There is no equivalent argument in the case of silicon.
The activation energies associated with the diffusion coefficient measured in
silicon and germanium vary from 4.8 to 5.2 eV and 3.0 to 3.2 eV, respectively.
The disagreement between quenching results (Sect.6.4.1) and self-diffusion results, particularly striking in the case of silicon, lead some to suggest that

self-diffusion could occur through various mechanisms: a split-vacancy mechanism (Sect.1.2.1) a divacancy mechanism or an "extended" defect (vacancy or interstitial) mechanism. Unfortunately, none of these mechanisms can be reconciled with the fact that the vacancy migrates at low temperature (Sect.6.4).

One of the arguments used to propose that self-diffusion is occuring in silicon through an extended defect, i.e., a defect which involves many lattice sites, is that the preexponential factor which is observed provides a very large entropy term as compared to the entropy value expected. But as we discussed in Sect.5.4.4, such a large entropy value is not at all in disagreement with a vacancy mechanism when a reasonable distortion around the vacancy (in agreement with direct experimental observation provided by electron paramagnetic resonance) is considered. The inconsistency which exists between quenching, annealing, and self-diffusion data, as already mentioned in Sect. 6.4 (Table 6.6), can be removed using arguments developed in this same section. The first argument consists in saying that the defects formed during the quenching process are not associations of vacancies with impurities, but are interstitial impurity complexes (in that case, the activation energy R associated with quenching is the formation enthalpy of the interstitial). The second argument, which seems the most reasonable one, is the following. The quenched defects are indeed vacancy-impurity complexes. However, interstitials, which are also present, but in a smaller concentration according to the conclusions we have derived in Sect.6.4.1, compete with the impurities to trap (and annihilate) vacancies. The consequence is that R will be a complicated function of the formation enthalpies for the vacancy and the interstitial, and of binding enthalpies between the vacancy, the interstitial and the impurities.

7.4.4 Substitutional Impurity Diffusion

We only consider here the case of the diffusion of Group III and V impurities. Because such impurities possess the same electronic bonding that the host atoms, it is natural to think that they diffuse through a mechanism identical to the self-diffusion mechanism, i.e., through the vacancy mechanism. Then the activation energy Q_i for the diffusion is the sum of the formation enthalpy H_F^V of the vacancy and of the migration enthalpy H_m^i of the impurity. Therefore, by substracting Q_i from the activation energy Q for self-diffusion, we obtain

$$Q_i - Q = H_m^i - H_m \quad . \tag{7.62}$$

Experimentally, as shown in Table 7.3, the values of Q_i are found to be smaller than Q in silicon and germanium. In silicon $H_m - H_m^i \simeq 1$ eV, which again is in disagreement with the fact that H_m^v is observed to be small (0.2-0.3 eV). The reason can be that, during diffusion, the charge state of the vacancy is different from the charge state of the vacancy-impurity (E-center) and we know that a change of charge state can induce a large change in the migration energy. Strong indication that indeed the charge states of the vacancy and of the impurity play an important role in the diffusion is found when we examine in detail a diffusion profile. The diffusion profile exhibits a characteristic kink (Fig.7.8) which indicates that it can be decomposed into the sum of two independent diffusion profiles having therefore diffusion coefficients associated with different activation energies [7.26].

Table 7.3. Activation energies [eV] for the diffusion of Group III and V impurities in silicon and germanium. References can be found in [7.27]

Element	Silicon	Germanium
B	3.7	4.5
Al	3.5	3.2
Ga	3.9	3.1
In	3.9	3.0
P	3.7	2.5
As	4.2	2.5
Sb	4.0	2.4

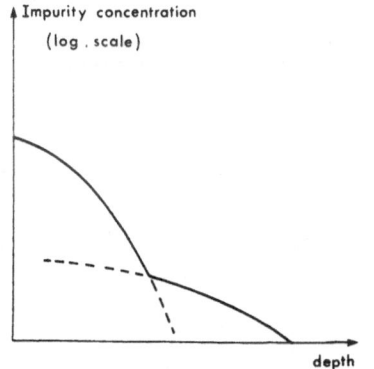

Fig.7.8. Schematic representation of a diffusion profile (variation of the impurity concentration with depth) exhibiting a kink

Because the diffusing impurity is also the doping impurity, the Fermi-level position varies with depth. It is reasonable to think that, at the depth the kink occurs, the Fermi-level position is such that it changes the charge state of one of the species (the impurity or the vacancy) involved in the diffusion. This charge-state change will induce a modification in the formation, migration, or binding enthalpy of this species and modify accordingly the diffusion coefficient. Moreover the preexponential factor D_0 will also be modified. Indeed, for the charge state which will correspond to an attractive Coulomb interaction between the impurity and the vacancy, D_0 will be multiplied by a factor NV_c, where N is the atomic density and V_c a critical volume. This critical volume expresses the fact that the vacancy, because it is very mobile at the temperature the diffusion occurs, does not have to be created as a neigbor of the impurity site, but within a critical distance r_c (such that for $r \simeq r_c$ the vacancy is attracted by the impurity). This critical radius r_c can be obtained by requiring the Coulomb attraction to be larger than the thermal energy for this attraction to occur:

$$\frac{e^2}{\varepsilon r_c} = kT \quad , \tag{7.63}$$

from which we estimate

$$V_c = \frac{4}{3} \pi \left(\frac{e^2}{\varepsilon kT}\right)^3 \quad . \tag{7.64}$$

Figure 7.9 shows that when the diffusion coefficient for self-diffusion,

$$D = D_0 \exp(-Q/kT) \quad , \tag{7.65}$$

is modified as discussed above, i.e., written as

$$D' = \frac{4}{3} \pi \left(\frac{e^2}{\varepsilon kT}\right)^3 ND_0 \exp\left(-\frac{Q}{kT}\right) \exp\left(+\frac{B}{kT}\right) \quad , \tag{7.66}$$

it becomes of the same order of magnitude as the diffusion coefficient for impurity diffusion. The experimental case considered here corresponds to phosphorus impurities diffusing into silicon, for which $Q_i = 3.5$ eV. The binding energy we take for next nearest neighbors is $B \simeq 0.5$ eV (Fig.6.3). A larger value of B, which would also include a possible elastic interaction (~ 0.5 eV for next nearest neighbors), i.e., taking $B \simeq 1$ eV, would ameliorate the fit.

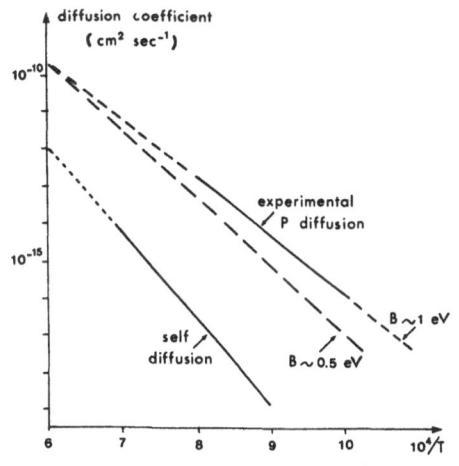

Fig.7.9. Arrhenius plots for self-diffusion and Phosphorus impurity diffusion in silicon. Full lines: experimental; dashed lines: from (7.66) with B = 0.5 eV and 1 eV

7.4.5 Interstitial Impurity Diffusion

Table 7.4 gives the activation energy associated with the diffusion for various impurities which are thought to diffuse through an interstitial mechanism. This table gives also the atomic masses and the ionic radii of the impurities.

Table 7.4. Activation energy for interstitial diffusion in silicon (Q_{Si}) and germanium (Q_{Ge}). The atomic mass A and the ionic radius R for the probable charge are also given.

Element	R [Å]	A [amu]	Q_{Si} [eV]	Q_{Ge} [eV]
Li	0.6	6.9	0.63 - 0.79	0.46 - 0.57
Na	0.95	23	0.72	-
K	1.33	39.1	0.76	-
H	2.08	1	0.48 - 0.56	0.38
He	-	4	0.58 - 1.26	0.61 - 0.70
Ag	-	107.9	1.60	-
Au	1.37	197	0.39	0.63
Zn	0.88	65.4	1.4	-
Cu	0.96	63.5	0.43	0.33
Fe	0.87	55.8	-	1.1

   The fact that impurities having a large atomic mass or a large ionic
radius can diffuse with an activation energy smaller than the activation
energy associated with other impurities of smaller mass or of smaller radius
indicates that both atomic mass and ionic radius (i.e., the elastic interaction) are not dominant parameters. A parameter which we can suspect to be
important is, once again, the charge state of the diffusing interstitial.
Enhancement of interstitial mobility due to charge-state changes has indeed
been observed. In n-type germanium the recombination of vacancy-interstitial
pairs, which occurs through the interstitial mobility [7.28] around 65 K,
can be induced at temperatures as low as 4 K when band-gap light illuminates
the sample [7.29]. In aluminum-doped silicon the self-interstitial is not
observed, even after 4 K electron irradiation, but aluminum interstitials
are produced (identified by electron paramagnetic resonance) [7.13]; this
implies that self-interstitials have to be mobile at the temperature of irradiation in order to be trapped on substitutional aluminum and to exchange
with them. Since this type of mobility occurs under ionization (irradiation
by energetic electrons produces electron-hole pairs), the mechanism involved
is probably an ionization mechanism such as the ones described in Sect.7.3.2
(the fact that interstitial mobility is still observed at 4 K suggests an
athermal mechanism).

References

Chapter 1

1.1   J. Bourgoin, M. Lannoo: *Point Defects in Semiconductors II*. Experimental
      Aspects, Springer Series in Solid State Sciences, Vol.35 (Springer, Berlin,
      Heidelberg, New York)
1.2   G. Burns, A.M. Glazer: *Space Groups for Solid State Scientists* (Academic,
      New York 1978)
1.3   M. Tinkham: *Group Theory and Quantum Mechanics* (McGraw-Hill, New York
      1964)
1.4   V. Heine: *Group Theory and Quantum Mechanics* (Pergamon, London 1960)
1.5   R.M. Hochstraner: *Molecular Aspects of Symmetry* (Benjamin, New York
      1966)
1.6   B.K. Vainshtein: *Modern Crystallography I*, Symmetry of Crystals, Methods
      of Structural Crystallography, Springer Series in Solid State Sciences,
      Vol.15 (Springer, Berlin, Heidelberg, New York 1981)
1.7   T.M. Morgan: *Proc. 11th Intern. Conf. on the Physics of Semiconductors*,
      Vol.2, ed. by M. Miasek (Elsevier, Amsterdam 1972) p.989
1.8   G.L. Bir, G.E. Pikus: *Symmetry and Strain Induced Effects in Semicon-
      ductors* (Wiley, New York 1974)
1.9   A.A. Kaplyanskii: Opt. Spectrosc. (USSR) *16*, 329 and 557 (1964)
1.10 A.M. Stoneham: *Theory of Defects in Solids* (Clarendon Press, Oxford
      1975) Chap.12

Chapter 2

2.1   G.H. Wannier: *Elements of Solid State Theory* (Cambridge University,
      Cambridge 1959)
2.2   J.M. Ziman: *Principles of the Theory of Solids* (Cambridge University,
      Cambridge 1969) p.151
2.3   O. Madelung: *Introduction to Solid-State Theory*, Springer Series in
      Solid-State Sciences, Vol.2 (Springer, Berlin, Heidelberg, New York
      1978)
2.4   W. Kohn: In *Solid State Physics*, Vol.5, ed. by F. Seitz, D. Turnbull
      (Academic, New York 1957) p.258
2.5   J. Callaway: *Energy Band Theory* (Academic, New York 1964)
2.6   A.M. Stoneham: *Theory of Defects in Solids* (Clarendon Press, Oxford
      1975)
2.7   S.T. Pantelides: Rev. Mod. Phys. *50*, 797 (1978)
2.8   F. Bassani, G. Iadonosi, B. Preziosi: Rept. Prog. Phys. *37*, 1099 (1974)
2.9   L. Pauling: *The Nature of the Chemical Bond* (Cornell University, New
      York 1960)

2.10 J.C. Phillips: *Bonds and Bands in Semiconductors* (Academic, New York 1973)
2.11 W.A. Harrisson: *Electronic Structure and the Properties of Solids, The Physics of the Chemical Bond* (Freeman, Reading 1980)
2.12 C.A. Coulson, L.B. Redei, D. Stocker: Proc. Roy. Soc. (London) *270*, 352 (1971);
     W.A. Harrisson: Phys. Rev. B*8*, 4487 (1973);
     M. Lannoo, J.N. Decarpigny: Phys. Rev. B*8*, 5704 (1973)
2.13 J.P. Walter, M.L. Cohen: Phys. Rev. B*4*, 1877 (1971)
2.14 P.O. Löwdin: J. Chem. Phys. *18*, 365 (1950)
2.15 J.M. Luttinger, W. Kohn: Phys. Rev. *97*, 869 (1955)
2.16 E.O. Kane: J. Phys. Chem. Solids *1*, 82 (1956)
2.17 R.F. Wallis, R. Herman, H.W. Milnes: J. Mol. Spectrosc. *4*, 51 (1960)
2.18 J.C. Slater: *Quantum Theory of Molecules and Solids*, Vol.1 (McGraw Hill, New York 1963) p.50
2.19 J.J. Hopfield: *Physics of Semiconductors* (Dunod, Paris 1964) p.725
2.20 S.M. Kogan, T.M. Lifshits: Phys. Status Solidi A*39*, 11 (1977)
2.21 E.E. Haller: Phys. Rev. Lett. *40*, 584 (1978)
2.22 B. Pajot, J. Fauppnen, R. Anttila: Solid State Commun. *31*, 759 (1979)
2.23 R.A. Faulkner: Phys. Rev. *175*, 991 (1968)
2.24 W.V. Smith, P.P. Sorokin, I.L. Gelles, G.J. Lasher: Phys. Rev. *115*, 1546 (1959)
2.25 J.C. Bourgoin, J. Krynicki, B. Blanchard: Phys. Status Solidi A*52*, 293 (1979)
2.26 T.J. Lee, T.C. McGill: J. Appl. Phys. *46*, 373 (1975)
2.27 W.G. Spitzer, M. Waldner: Phys. Rev. Lett. *14*, 223 (1965)
2.28 P.J. Dean: Phys. Rev. A*139*, 588 (1965)
2.29 J.J. Hopfield, D.G. Thomas, M. Gershinzon: Phys. Rev. Lett. *10*, 162 (1963)
2.30 J.S. Prener: J. Chem. Phys. *25*, 1294 (1956)

Chapter 3

3.1 W.A. Harrisson: *Electronic Structure and the Properties of Solids, The Physics of the Chemical Bond* (Freeman, New York 1980) Chap.2
3.2 L. Salem: *The Molecular Orbital Theory of Conjugated Systems* (Benjamin, New York 1972)
3.3 J.C. Slater: *Quantum Theory of Atomic Structure*, Vol.1 (McGraw Hill, New York 1960) p.206
3.4 J.C. Slater, G.J. Koster: Phys. Rev. *94*, 1498 (1954)
3.5 E. Kauffer, P. Pécheur, M. Gerl: J. Phys. C*9*, 2319 (1976)
3.6 J.P. Walter, M.L. Cohen: Phys. Rev. B*4*, 1877 (1971)
3.7 W.A. Harrisson: Phys. Rev. B*8*, 4487 (1973);
     M. Lannoo, J.N. Decarpigny: Phys. Rev. B*8*, 5704 (1973)
3.8 G. Leman, J. Friedel: J. Appl. Phys. Suppl. *33*, 281 (1962)
3.9 D. Weaire, M.F. Thorpe: Phys. Rev. B*4*, 2508 (1971)
3.10 F. Cyrot-Lackmann: J. Phys. C*5*, 300 (1972)
3.11 G.D. Watkins: J. Phys. Soc. Jpn. (Suppl. II) *18*, 22 (1963); also *Radiation Effects in Semiconductors* (Dunod, Paris 1964) p.97
3.12 T. Yamaguchi: J. Phys. Soc. Jpn. *17*, 1359 (1962)
3.13 W.A. Harrison: The Physics of Solid State Chemistry, Festkörperprobleme XVII (Springer Tracts C, Berlin, Springer 1977)
3.14 H.P. Hjalmarson, P. Vogl, D.J. Wolford, J.D. Dow: Phys. Rev. Lett. *44*, 810 (1980)
3.15 E.N. Economou: *Green's Functions in Quantum Physics*, Springer Series in Solid-State Sciences, Vol.7, (Springer, Berlin, Heidelberg, New York 1979)

3.16 S.T. Pantelides: Rev. Mod. Phys. *50*, 797 (1978)
3.17 J. Callaway: J. Math. Phys. *5*, 783 (1964)
3.18 A. Messiah: *Mécanique Quantique*, Vol.2 (Dunod, Paris 1960) p.702
3.19 I.M. Lifshitz, L.N. Rosenzweig: Izv. Akad. Nauk. SSSR Ser. Fiz. *12*, 667
     (1948);
     I.M. Lifshitz: Nuovo Cimento Suppl. *3*, 716 (1956);
     J. Friedel: Ann. Phys. *9*, 158 (1954)
3.20 G.J. Koster, J.C. Slater: Phys. Rev. *95*, 1167 (1954)
3.21 F. Cyrot-Lackmann: Adv. Phys. *16*, 393 (1967); J. Phys. Chem. Solids *29*,
     1235 (1968)
3.22 R.C. Gordon: J. Math. Phys. *9*, 655 (1968)
3.23 R. Haydock, V. Heine, M.J. Kelly: J. Phys. C*5*, 2845 (1972)
3.24 M. Lannoo: Ann. Phys. Paris *3*, 391 (1968);
     M. Lannoo, P. Lenglart: J. Phys. Chem. Solids *30*, 2409 (1968)
3.25 E. Kauffer, P. Pêcheur, M. Gerl: Phys. Rev. B*15*, 4107 (1977)
3.26 P. Pêcheur, E. Kauffer, M. Gerl: Phys. Rev. B*14*, 4521 (1976)
     F. Bernholc, S.T. Pantelides: Phys. Rev. B*18*, 1780 (1978)
3.27 E. Kauffer, P. Pêcheur, M. Gerl: Rev. Phys. Appl. *15*, 849 (1980)

Chapter 4

4.1 W.J. Hunt, P.J. Hay, W.A. Goddard III: J. Chem. Phys. *57*, 738 (1972)
4.2 P. Hohenberg, W. Kohn: Phys. Rev. *136*, 864 (1964)
4.3 J.C. Slater: In *Quantum Theory of Molecules and Solids*, Vol.4 (McGraw-
    Hill, New York 1974) p.56 and 293
4.4 M.A. Whitehead: In *Sigma Molecular Orbital Theory*, ed. by O. Sinanoğlu,
    K.B. Wiberg (Yale University Press, New Haven, 1970) p.49
4.5 J.N. Decarpigny, M. Lannoo: Phys. Rev. B*14*, 538 (1976)
4.6 M. Lannoo: Phys. Rev. *10*, 2544 (1974)
4.7 F.D.M. Haldane, P.W. Anderson: Phys. Rev. B*13*, 2553 (1976)
4.8 G.A. Baraff, E.O. Kane, M. Schlüter: Phys. Rev. Lett. *43*, 956 (1979)
4.9 P. Pêcheur, E. Kauffer, M. Gerl: In *Lattice Defects in Semiconductors
    1978* (Institute of Physics, London 1979) Conf. Ser. 46, p.174
4.10 J.A. Pople, G.A. Segal: J. Chem. Phys. *44*, 3289 (1966)
4.11 A.H. Harker, F.P. Larkins: J. Phys. C*12*, 2497 (1979)
4.12 A. Mainwood: J. Phys. C*11*, 2703 (1978)
4.13 R.P. Messmer, G.D. Watkins: In *Radiation Damage in Semiconductors*, ed.
     by J.W. Corbett, G.D. Watkins (Gordon and Breach, New York 1971) p.23
4.14 R.P. Messmer: Chem. Phys. Lett. *11*, 589 (1971)
4.15 R. Hoffman: J. Chem. Phys. *39*, 1397 (1963)
4.16 R.S. Mulliken: J. Chem. Phys. *23*, 1833, 1841, 2238, 2343 (1955)
4.17 M. Astier, M. Pottier, J.C. Bourgoin: Phys. Rev. B*19*, 5265 (1979)
4.18 C.C.J. Roothaan: Rev. Mod. Phys. *23*, 69 (1957)
4.19 W. Kohn, L.J. Sham: Phys. Rev. *140*, A 1133, (1965)
4.20 G.A. Baraff, M. Schlüter: Phys. Rev. Lett. *41*, 892 (1978)
4.21 G.A. Baraff, M. Schlüter: Phys. Rev. B*19*, 4965 (1979)
4.22 S.G. Louie, M. Schlüter, J.R. Chelikowsky, M.L. Cohen: Phys. Rev. B*13*,
     1654 (1976)
4.23 E. Kauffer, P. Pêcheur, M. Gerl: J. Phys. C*9*, 2319 (1976);
     Phys. Rev. B*15*, 4107 (1977);
     P. Pêcheur, E. Kauffer, M. Gerl: Phys. Rev. B*14*, 4521 (1976)
4.24 J. Bernholc, S.T. Pantelides: Phys. Rev. B*18*, 1780 (1978)
4.25 J. Bernholc, N.O. Lipari, S.T. Pantelides: Phys. Rev. Lett. *41*, 895
     (1978)
4.26 C.O. Rodriguez, S. Brand, M. James: *Defects and Radiation Effects in
     Semiconductors* (Institute of Physics, London 1979) Conf. Ser. 46, p.193

4.27 G. Srinivasan: Phys. Rev. *178*, 1244 (1969)
4.28 M. Lannoo, G. Allan: Solid State Commun. *33*, 293 (1980)
4.29 J.C. Slater: *Quantum Theory of Molecules and Solids*, Vol.1 (McGraw-Hill, New York 1963) p.285
4.30 G.D. Watkins: J. Phys. Soc. Jpn. (Suppl. 2) *18*, 22 (1963); *Radiation Effects in Semiconductors*, ed. by P. Baruch (Dunod, Paris 1964) p.97
4.31 C.A. Coulson, M.J. Kearsley: Proc. Roy. Soc. London A*241*, 433 (1957)
4.32 C.D. Clark, R.W. Ditchburn, H.E. Dyer: Proc. Roy. Soc. London A*234*, 363 (1956); A*237*, 75 (1956)
4.33 R.S. Mulliken: Phys. Rev. *43*, 279 (1933)
4.34 F.P. Larkins: J. Phys. Chem. Solids *32*, 965 (1971)
4.35 C.A. Coulson, F.P. Larkins: J. Phys. Chem. Solids *32*, 2245 (1971)
4.36 G.T. Surratt, W.A. Goddard III: Solid State Commun. *22*, 413 (1977)
4.37 F.P. Larkins: J. Phys. Chem. Solids *32*, 2245 (1971)
4.38 W.A. Harrisson: In *The Physics of Solid State Chemistry*, ed. by J. Treusch, Festkörperprobleme 17 (Vieweg and Son, Braunschweig, 1977)
4.39 G.D. Watkins, R.P. Messmer: Phys. Rev. Lett. *32*, 1244 (1974)

Chapter 5

5.1 G.K. Horton, A.A. Maradudin (eds.): *Dynamical Properties of Solids* (North Holland, Amsterdam 1974)
5.2 A.A. Maradudin, E.W. Montroll, G.H. Weiss: "Theory of Lattice Dynamics in the Harmonic Approximation", in *Solid State Physics*, ed. by F. Seitz, D. Turnbull (Academic, New York 1963) Suppl.3
5.3 H. Bilz, W. Kress: *Phonon Dispersion Relations in Insulators*, Springer Series in Solid-State Sciences, Vol.10 (Springer, Berlin, Heidelberg, New York 1979)
5.4 M. Born, K. Huang: *Dynamical Theory of Crystal Lattices* (Clarendon, Oxford 1968)
5.5 M. Lannoo: J. Phys. Paris *40*, 461 (1979)
5.6 D. Weaire, R. Alben: Phys. Rev. *29*, 1505 (1972)
5.7 R.H. Lyddane, R. Sachs, E. Teller: Phys. Rev. *59*, 673 (1941)
5.8 A.A. Maradudin: In *Solid State Physics*, Vols. 18 and 19, ed. by F. Seitz, D. Turnbull (Academic, New York 1966)
5.9 R.A. Swalin: *Thermodynamics of Solids* (J. Wiley and Sons, New York 1962)
5.10 L. Dobrzynski, J. Friedel: Surf. Sci. *12*, 649 (1968);
B. Djafari-Rouhani, L. Dobrzynski, G. Allan: Surf. Sci. *55*, 663 (1976)
5.11 M. Lannoo, J.C. Bourgoin: Solid State Commun. *32*, 913 (1979)
5.12 A. Mainwood: J. Phys. C*11*, 2703 (1978)
5.13 M. Astier, M. Pottier, J.C. Bourgoin: Phys. Rev. B*19*, 5265 (1979)
5.14 G. Allan, M. Lannoo: Proc. Int. Conf. on Defects in Semiconductors, Oiso, Japan, 1980 (To be published)

Chapter 6

6.1 R.A. Swalin: *Thermodynamics of Solids* (Wiley, New York 1962)
6.2 A.D. Franklin: In *Point Defects in Solids*, Vol.1, ed. by J.H. Crawford, Jr., L.M. Slifkin (Plenum, New York 1972) Chap.1
6.3 R.A. Swalin: J. Phys. Chem. Solids *18*, 290 (1961)
6.4 A. Seeger, M.L. Swanson: In *Lattice Defects in Semiconductors*, ed. by R.R. Hasiguti (University of Tokyo, Tokyo 1968) p.93
6.5 K.H. Benneman: Phys. Rev. A*137*, 1497 (1965)
6.6 C.J. Huang, L.A.K. Watt: Phys. Rev. *171*, 958 (1961)
6.7 R.A. Swalin: Acta Metall. *7*, 736 (1959)

6.8 T.N. Morgan: *Proc. 11th Intern. Conf. on the Physics of Semiconductors*, Vol.2, ed. by M. Miasek (Elsevier, Amsterdam 1972) p.989
6.9 J.D. Eshelby: In *Solid State Physics*, Vol.3, ed. by F. Seitz, D. Turnbull (Academic, New York 1956) p.79
6.10 R.E. Howard, A.B. Lidiard: Rep. Prog. Phys. *27*, 161 (1964)
6.11 H.C. Casey, Jr.: In *Atomic Diffusion in Semiconductors*, ed. by D. Shaw (Plenum, New York 1973) Chap.6
6.12 L. Estner, W. Kamprath: Phys. Status Solidi *22*, 541 (1967)
6.13 A. Hiraki, T. Suita: J. Phys. Soc. Jpn. (Suppl. 3) *18*, 254 (1963)
6.14 B. Samuelsson: Ark. Fys. *35*, 321 (1967)
6.15 H. Letaw, Jr.: J. Phys. Chem. Solids *1*, 100 (1956)
6.16 R.A. Logan: Phys. Rev. *101*, 1455 (1956)
6.17 A. Hiraki: J. Phys. Soc. Jpn. *21*, 34 (1968)
6.18 S. Mayburg, L. Rotondi: Phys. Rev. *91*, 1015 (1953)
6.19 S. Ishino, F. Makazawa, R.R. Hasiguti: J. Phys. Soc. Jpn. *20*, 817 (1965)
6.20 A. Scholz: Phys. Status Solidi *3*, 42 (1963)
6.21 R.E. Whan: Phys. Rev. A*140*, 690 (1965)
6.22 G.D. Watkins: J. Phys. Soc. Jpn. (Suppl. 2) *18*, 22 (1963)
6.23 H. Reiss, C.S. Fuller, F.J. Morin: Bell Syst. Tech. J. *35*, 535 (1956)

Chapter 7

7.1 H.R. Glyde: Rev. Mod. Phys. *39*, 2 (1967)
7.2 E. Wigner: Phys. Rev. *40*, 749 (1932); Z. Phys. Chem. Abt. B*19*, 203 (1932)
7.3 S.A. Rice: Phys. Rev. *112*, 804 (1958)
7.4 H.R. Glyde: J. Phys. Chem. Solids *28*, 2061 (1967)
7.5 J.A. Van Vechten: Phys. Rev. B*12*, 1247 (1975)
7.6 R.A. Swalin: J. Phys. Chem. Solids *18*, 290 (1961)
7.7 A. Seeger, M.L. Swanson: In *Lattice Defects in Semiconductors*, ed. by R.R. Hasiguti (University of Tokyo, Tokyo 1968) p.93
7.8 G.M. de Mussari, L. Gabba, J. Guisano, G. Mambriane: Phys. Status Solidi A*34*, 455 (1976)
7.9 R.R. Hasiguti: J. Phys. Soc. Jpn. *21*, 1927 (1966)
7.10 C. Weigel. J.W. Corbett: Z. Phys. B*23*, 233 (1976)
7.11 T. Soma, M. Saeke, A. Morita: J. Phys. Soc. Jpn. *35*, 146 (1973)
7.12 G.D. Watkins: J. Phys. Soc. Jpn. (Suppl. II) *18*, 22 (1963)
7.13 G.D. Watkins: In *Radiation Damage in Semiconductors*, ed. by P. Baruch (Paris, Dunod 1964) p.97
7.14 R.E. Whan: Phys. Rev. *140*, A690 (1955)
7.15 C.D. Clark, J. Walker: Proc. Roy. Soc. London A*334*, 241 (1973)
7.16 K. Weiser: Phys. Rev. *126*, 1427 (1962)
7.17 N.F. Mott, J. Littleton: Trans. Faraday Soc. *34*, 485 (1948)
7.18 L. Pauling: *The Nature of the Chemical Bond* (Cornell University, New York 1960)
7.19 J.C. Bourgoin, J.W. Corbett: Radiat. Eff. *36*, 157 (1978)
7.20 M. Lax: J. Phys. Chem. Solids *8*, 66 (1952)
7.21 H. Grummel, M. Lax: Phys. Rev. *97*, 1469 (1955); I. Toyazawa: Prog. Theor. Phys. *13*, 160 (1955)
7.22 J.R. Manning: *Diffusion Kinetics for Atoms in Crystals* (Van Nostrand, Princeton 1968); C.P. Flynn: *Point Defects and Diffusion* (Clarendon, Oxford 1972) Y. Adda, J. Philibert: *La Diffusion dans les Solides* (Presses Universitaires de France 1966)
7.23 D. Shaw (ed.): *Atomic Diffusion in Semiconductors* (Plenum, New York) 1973);

H.C. Casey, G.L. Pearson: In *Point Defects in Solids*, Vol.2, ed. by J.H. Crawford, L.M. Slifkin (Plenum, New York 1975) Chap.2; A.F. Willoughby: Rep. Prog. Phys. **41**, 1665 (1978); D. Shaw: Phys. Status Solidi B*72*, 11 (1975)

7.24 T.H. Yeh: In *Atomic Diffusion in Semiconductors*, ed. by D. Shaw (Plenum, New York 1973) Chap.4

7.25 A.G. Tweet: Phys. Rev. *106*, 221 (1957)

7.26 R. Francis, P.S. Dobson: *Lattice Defects in Semiconductors*, Conf. Ser. 23 (Institute of Physics, London 1975) p.545

7.27 S.M. Hu: In *Atomic Diffusion in Semiconductors*, ed. by D. Shaw (Plenum, New York 1973) Chap.5

7.28 J.W. Mac Kay, E.E. Klontz: J. Appl. Phys. *30*, 1269 (1959)

7.29 I. Arimura, J.W. Mac Kay: In *Radiation Effects in Semiconductors*, ed. by F.L. Vook (Plenum, New York 1968) p.186

Subject Index

Absorption
  in diamond 142
  optical 60,161,239
A center 7,229
Acoustic
  branch 162,164
  mode 160
  wave propagation 161
  wave velocity 159
Aggregate 2,8,201
Aluminum
  interstitial 244
  substitutional 244
Amorphous 2,3,75
Annealing 204,215,228,240
Annihilation 229
Antibondig
  energy 233
  H_2 orbital 133,136
  impurity state 87,88
  interstitial state 85,86
  orbital 81,82,90
  state 75,77-79,88,89
Antimony 63
  binding energy in Si 82
Antiresonance 98,101
Antisite 2,208
Arsenic 63
  binding energy in Si 62
Association 1,229
Athermal
  ionization enhanced migration
    234,236

migration 235,244
Atomic mass 228,243,244

Backscattering 239
Ballistic model 227
Band
  conduction 45,61,70,75,79,82,83,86
  gap 36,38,39,55,60,65,68,119,195,244
  structure 70,71,73-77,124
  valence 38,52,70,75,77-79,82,83
Base
  perturbative potential 130
  pseudopotential 130
Basis function 11,12
Bethe lattice 80,108,110
Binding
  Boron in diamond 66
  energy 53-55,57,62,65,215,218
  free exciton 57
  pair 58,208
  state 47,61,75
Bloch
  function 44,48,50,126
  state 39,97
  sum 160
  theorem 10,158,160
Bohr
  acceptor 56
  donor 56,58
  radius 39,54
Boltzmann statistics 222
Bond
  angle 3,124

256

Energy (continued)

promotion (sp) 72,73,86,87-89,192

release mechnism 236

total 113,124,125

Enhanced migration

athermal ionization 234,236

ionization 232,233

Enthalpy 195,212,214

charge-state dependence 9

formation 191,211,213,215,224,233

impurity binding 242

impurity formation 242

impurity migration 242

interstitial formation 192,240

interstitial migration 228,237

migration 215,223,224,227-230,233

vacancy formation 182,192-194,230, 233,237

vacancy migration 87,185,228

Entropy 23,156,162,188,195,196,198, 214,233,237

charge-state dependence 9

determination 189

formation 180,189,211,213

interstitial formation 212

ionization 189

self-diffusion 184,187,240

vacancy 180,186,190

vacancy formation 155,185,189,212

vacancy migration 185,187,189

vibrational 155,180ff.,223,224

Equilibrium configuration 231,236

Exchange

correlation potential 114,130

potential 114

term 114,138

Expansion 213

Extended Hückel theory 123-125

Fermi level 199,200,229,232-234,242

Fick's law 237,238

Floating zone 7

Forbidden energy gap 36,38,39,60,65, 83,96,107,111,116,119,195,244

Force constant 156,158,159,167,168, 173,177,222,223,225

defect 176

local 165

matrix 156,157,162

model 160,183

vacancy 184-187

Formation

energy 9,224,231

enthalpy 191,211,213,215,224

Fourier transform 39,40,48

Free energy 180,189,195,197,199,200, 205,209,211,215,224

binding 201,204,211

formation 200,201,206

ionization 200

migration 223

pair formation 201,202

Frenkel pair 1,210

formation enthalpy 191

Gallium arsenide 61

donor binding energy 62

effective mass 62

Gallium phosphide 91

Gap 36,38,39,55,60,65,68,83,91,96, 107,111,116,117,119,121,123,195,244

Generalized phase shift 98

Germanium

atomic mass 228

bond energy 37,193,194

cohesive energy 193

conduction band 44

Debye temperature 228

dielectric constant 117

diffusion of group 3 impurity 241

diffusion of group 5 impurity 241

donor binding energy 62

effective Coulomb energy 117

effective mass 52,62

G. Leibfried, N. Breuer

Point Defects in Metals I

Introduction to the Theory

1978. 138 figures, 22 tables. XIV, 342 pages
(Springer Tracts in Modern Physics, Volume 81)
ISBN 3-540-08375-8

Contents:
Introduction and survey. – Harmonic approximation and linear response (Green's function) of an arbitrary system. – Lattice theory. – Continuum theory. Transition from lattice to continuum theory. – Statics and dynamics of simple point defects. – Scattering of neutrons and X-rays by crystals. – Probability, distributions and statistics. – Properties of crystals with defects in small concentration. – Appendix.

Point Defects in Metals II

**Dynamical Properties
and Diffusion Controlled Reactions**

1980. 91 figures. 6 tables. X, 262 pages
(Springer Tracts in Modern Physics, Volume 87)
ISBN 3-540-09623-X

Springer-Verlag
Berlin
Heidelberg
New York

Contents:
P. H. Dederichs, R. Zeller: Dynamical Properties of Point Defects in Metals
K. Schroeder: Theory of Diffusion Controlled Reactions of Point Defects in Metals

Condensed Matter

Zeitschrift für Physik B

Unter Mitwirkung der
Deutschen Physikalischen Gesellschaft
EPS Europhysics Journal
ISSN 0340-224X   Title No. 257

Editor in Chief: H. Horner, Heidelberg

Editorial Board: H. Bilz, Stuttgart; W. Brenig, Garching;
W. Buckel, Karlsruhe; M. Campagna, Jülich;
J. Christiansen, Erlangen; R. A. Cowley, Edinburgh;
W. Klose, Karlsruhe; H. C. Siegmann, Zürich; T. Springer,
Grenoble; P. Szépfalusy, Budapest; H. Thomas, Basel;
J. Zittartz, Köln

Zeitschrift für Physik appears in three parts:
A: Atoms and Nuclei
B: Condensed Matter
C: Particles and Fields
Each part may be ordered separately.

Coordinating Editor of Section A, B and C:
O. Haxel, Heidelberg

Zeitschrift für Physik B
Condensed Matter

Physics of Condensed Matter
● Physical properties of crystalline, disordered and
  amorphous solids
● Classical and quantum-fluids
● Topics of molecular physics related to the physics
  of condensed matter
● Polymers

General Physics
● Quantum Physics
● Statistical physics, nonequilibrium and cooperative
  phenomena

Language: English

Subscription information and/or sample copies are available from
your bookseller or directly from
Springer-Verlag, Journal Promotion Dept., P.O. Box 105 280,
D-6900 Heidelberg, FRG

Springer-Verlag
Berlin
Heidelberg
New York